PRESENTED TO
Josephine LaCoe
SWE
BY THE
LEHIGH VALLEY SECTION
SOCIETY OF WOMEN ENGINEERS
MAY 5, 2013

Changing Our World

Changing Our World

True Stories of Women Engineers

Sybil E. Hatch

ISBN (hardcover) 0-7844-0841-6
ISBN (paperback) 0-7844-0835-1

Library of Congress Cataloging-in-Publication Data

Hatch, Sybil E.
Changing our world : true stories of women engineers / Sybil E. Hatch.
p. cm.
ISBN 0-7844-0841-6 (hardcover : alk. paper)—ISBN 0-7844-0835-1 (softcover : alk. paper)
1. Women engineers. I. Title.
TA157.H4155 2006
620.0082—dc22 2005033884

Book Design: Lisa Elliot/Elysium
Printed in Canada

Published by the American Society of Civil Engineers
1801 Alexander Bell Drive
Reston, Virginia 20191
www.pubs.asce.org

Contents

A Woman Engineer's Perspective

Science, engineering, and technology are all about creativity. What's more exciting than understanding the world around us? And then to use that understanding to communicate ideas and help shape our society and world? We do this through art and we do it through the tools—technology—that we create.

To me, science is much broader than biology, chemistry, physics, and their varied subsets. It encompasses all our attempts to question, wonder about, and understand the world. It includes the social, physiological, political, and behavioral aspects of our lives and how we interact with our environment.

Technology is not just contraptions made of nuts and bolts, or wood and metal, or chemicals and electricity. The best technologies can be very simple and can even be networks or structures. They reflect who we are and what we want to accomplish.

It's of critical importance, then, that we include more women and people of color in the fields of science, engineering, and technology. Why? Because the tools we create depend on the perspective of the people who create them!

Let me give an example. When I was an astronaut, women were issued the same flight clothing with dimensions designed by and for men. I'm tall and thin. I literally floated around in the space suit that fit my height! I have long thin fingers, but had to use space gloves designed for men's wide hands. We women astronauts at the time "made do," but think about it: would you try to stuff a man into clothing designed for a woman's shape and size?

Engineers play a critical role in creating technological tools. As humans, our thoughts naturally range between abstract and concrete concepts. What's wonderful about an engineering education is that it helps to put order into the thought process: engineering brings the abstract into focus.

Chemical engineering encompasses a wide range of materials, and overlaps with many different numerical systems. My background in chemical engineering allowed me to develop various ways to describe aspects of our world in quantitative ways—through mathematics and science.

Engineering also enabled me to see the world in new and qualitative ways, to make valid analogies between events and ideas from known scientific perspectives, and to know to ask questions when a certain perspective flies in the face of social, cultural, or physical reality.

Science and engineering from a woman's perspective? Our world is based on who we are and who participates! Through science and engineering, there are limitless opportunities for girls and women to create the tools we need to shape our future!

—Dr. Mae Carol Jemison
President, BioSentient Corporation

SHE REACHED FOR THE STARS—AND TOUCHED THEM

As a young girl growing up in Chicago, Illinois, chemical engineer **MAE CAROL JEMISON** (b.1956) looked up into the night sky and marveled at the stars, the constellations, and the universe. She told those who asked her what she wanted to be when she grew up, "a scientist."

A scientist she became! And a chemical engineer, an Area Peace Corps Medical Officer, a physician, an astronaut, a science educator-advocate, a dancer, an author, a speaker, an entrepreneur, and a mentor to hundreds of girls and boys. Mae Jemison's career and life reflect her determination to make the world a better place for all people.

Mae graduated from high school at 16, and went to Stanford University in California to study the new field of biomedical engineering. She graduated in four years with two degrees: one in chemical engineering and the other in African and Afro-American studies. She also took several graduate-level classes in biomedical engineering, and even became conversant in Swahili!

As she notes in her autobiography, *Find Where the Wind Goes: Moments from my Life,* "trying, risking, and being willing to put in the effort to accomplish the task led the way to one of the most positive and enabling experiences in my entire life."

Next stop: Cornell University Medical College where Mae obtained her medical degree. Mae then served in the Peace Corps where she was responsible for the health of all the Peace Corps volunteers and other Americans in Sierra Leone and Liberia. Upon her return, she practiced medicine in Los Angeles.

Ever since she was a child, Mae had always wanted to go into space. She applied to become an astronaut and was accepted in 1987. In 1992, she flew on the Space Shuttle *Endeavor* as the science mission specialist.

Afterward, she founded The Jemison Group, Inc., a firm that focuses on integrating science and technology into the everyday lives of people around the world. Her company developed solar thermal electricity generation systems for developing countries. It is also working on using a satellite-based telecommunications system called ALAFIYA (a Yoruba word meaning "good health") to facilitate health care in remote regions of West Africa.

In 1994, Mae founded The Earth We Share™ (TEWS), to help students and teachers learn about science, technology, and our world in a fun, meaningful way—to stimulate and maintain student interest in science.

Recently, Mae has returned to chemical and biomedical engineering by founding BioSentient Corporation. Mae is guiding the company in the design and development of "MobileMe," a garment that contains sensors to measure a person's heart rate, respiratory rate, pulse, temperature, and skin conductance. Originally designed to control motion sickness, MobileMe can now be used for a much wider range of human health and performance uses.

MAE JEMISON WAS THE FIRST AFRICAN-AMERICAN WOMAN IN SPACE. WHILE ABOARD, SHE CONDUCTED MORE THAN 44 DIFFERENT BIOMEDICAL EXPERIMENTS—INCLUDING HOW FROG EMBRYOS DEVELOPED IN WEIGHTLESSNESS.

What's next in Mae's professional life? "There are infinite possibilities," Mae says. "You'll just have to wait and see!"

Women Engineers Unite

The Society of Women Engineers (SWE) was officially founded on May 27, 1950, at its first "national convention," held at the Green Engineering Camp of the Cooper Union in New Jersey. About 50 women engineers from New York City, Philadelphia, Washington D.C., and Boston attended and elected **BEATRICE HICKS** (see page 162) as the first president.

Today, SWE has more than 19,000 student and professional members. SWE is a driving force that establishes engineering as a highly desirable career choice for women. SWE empowers women to succeed and advance in those aspirations and be recognized for their life-changing contributions and achievements as engineers and leaders.

THE BIRTH OF A VITAL PROJECT

When I was in high school in Lexington, Kentucky, engineering hadn't even entered my mind. I was an artist and dancer.

Then, an inspiring lecture by a University of Kentucky professor changed my world. He demonstrated how my love of art was easily incorporated into engineering. He showed me how engineering helped people—something I was very keen on doing. He talked about how engineering desperately needed women to provide a new perspective and to improve the profession. I was sold!

I talked with some people about engineering, but the response was less than encouraging. My guidance counselor said, "You have no aptitude for engineering." My math teacher said, "You'll flunk out." And worst of all, my grandmother said, "Isn't that a man's job?"

Luckily, I listened to the words of my mother instead. She said, "You can do anything and don't accept that it can't be done!" Discouragements only mean opportunities for you to show the world what you can really do!

My mother's words have echoed true throughout my career, and they led me to fulfill another dream: publish a book that would tell the stories of women engineers who could serve as role models to girls.

The ASCE Task Force Committee on Women in Civil Engineering worked for two years on the project. But it wasn't until I became ASCE President in 2003 that I had the opportunity to discuss the Task Force Committee's work with other major engineering society presidents, who, for the first time in history, were all women!

These discussions led to the birth of the Extraordinary Women Engineers Project Coalition, which has supported work on this book and other tools that will inform girls, guidance counselors, teachers, and parents about why engineering is such an exciting career—and how engineers improve the quality of life of all people.

Read more about **PAT GALLOWAY'S** engineering career on page 202.

I'm proud to be an engineer. I'm also proud to serve as the chair of this project. I hope that you enjoy reading the stories in this book as much as I have, and that they will inspire you or someone you know to choose engineering as a rewarding career!

—Patricia D. Galloway, Ph.D., P.E.
Civil Engineer
Chair, Steering Committee
Extraordinary Women Engineers Project

A Portal to Your Future

The first time I looked through this special book, I was amazed at the influence that women engineers have had on making the world a better place. However, by the end of the book, I realized that their amazing stories are just a small part of what makes this book so important. It is the shared voice of 238 women engineers—telling us that engineering is an exciting and rewarding career—that I find so impressive.

Read more about **SUE SKEMP'S** engineering career on page 163.

As these women tell us, engineering is not just one thing. Rather, it's a learned methodology combined with a desire to explore and innovate. Many of the women engineers in this book set out only to do a job or answer a question or solve a problem. However, in doing so, they changed the world! The opportunities are bounded only by human creativity—which, in my opinion, is limitless!

Though this network of women and experiences was not available when I made my decision to step into engineering, I envisioned a career that was exciting, challenging, and very rewarding—and it has turned out to be just that and more! Today, many more opportunities and career paths exist than 25 years ago when I graduated. Women today are gaining recognition for their achievements in engineering, their profession, and their careers, and contributions to society. I continue to learn from the ever-expanding network, and to explore new frontiers.

How important is an engineering background to our future? Studies already indicate that young people today will have more than one career in their lives, and that their careers will span business, government, and education. An engineering degree provides an excellent basis to analyze a situation and come up with a solution.

I hope you are as touched as I am by the excitement, enthusiasm, and accomplishments of the women engineers in this book, and use their experiences to open new vistas for you. Even though you may never meet them, they can serve as mentors and guides for you today in navigating your future. I'm proud to be part of this group and to be part of the engineering community that strives to make our world a better place.

— Susan H. Skemp
Mechanical Engineer
Chair, Advisory Committee
Extraordinary Women Engineers Project

Women Engineers Rule

The year 2003 was an amazing one in engineering. For the first time ever, the large engineering society presidents were all women—proving that women are up to the challenge of leading the engineering profession! From left to right:

- **DIANNE DORLAND,** first woman president of the American Society of Chemical Engineers.
- **SUSAN SKEMP,** second woman president of the American Society of Mechanical Engineers.
- **TERESA HELMLINGER,** first woman president of the National Society of Professional Engineers.
- **LEEARL BRYANT,** first woman president of the Institute of Electrical and Electronics Engineers —USA.
- **PATRICIA GALLOWAY,** first woman president of the American Society of Civil Engineers.

CHANGING
OUR WORLD

Imagine the Possibilities!

"That's a wrap!" the creative director yelled out after several long moments of silence. All of a sudden, everyone went crazy, jumping up and down, screaming, hugging, rolling on the floor and laughing!

Read more about **SHALINI GOVIL-PAI'S** engineering career on page 108.

"It was wonderful madness!" says computer engineer Shalini Govil-Pai. Shalini was technical director for the blockbuster animated movie A Bug's Life. *Pixar Animation Studios used its own breakthrough software—which Shalini helped develop—for realistic 3-D modeling, animation, and lighting.*

Shalini's team worked with artists and animators to make sure that the software did what the artists needed it to do. During the final weeks, almost everyone brought sleeping bags and toothbrushes to the office so that they could work

late into the night, go to sleep in their cubicles, and wake up ready to start again. "It was like a big slumber party, except that we were all working," says Shalini.

The very last day of production arrived. They still had 10 shots left to go—and each shot could take hours! But they did it! "It was music to our ears when the creative director approved the shots for film! I can't tell you how happy we were! It was amazing!"

Fast forward to the beautiful Orpheum Theater in downtown San Francisco for a private showing of A Bug's Life. *"It was the first time my team had seen the completed film with everything in place . . . the music, the voices, all the sound. We had tears of joy in our eyes as the movie ended. What an incredible night! I'll always remember it," says Shalini.*

Engineering Is All Around Us

Imagine seeing your own work on the big screen! Or seeing your own work in print. Better yet, imagine seeing your work every time you take a look around you! Think of the satisfaction you'd have knowing that you helped create the iPod in your backpack. The potato chips in your lunch. The car your family drives or the safe roads they drive on. The water from your tap. The electricity from your outlet.

Engineering creativity is all around us. The work of engineers has a tremendous influence—every day—on improving the health and well-being of people, society, and the environment.

This book is about the passion and commitment that women engineers bring to their own lives, their families, and to people around the globe. The women in this book come from different walks of life, different places, and even different decades. But they have one thing in common: they wanted to give back to society and make the world a better place to live.

"The future belongs to those who believe in the beauty of their dreams."

Eleanor Roosevelt
First Lady, Human Rights Advocate, Author, and Humanitarian (1884–1962)

What Is Engineering, Anyway?

Even though engineers' efforts are all around us, the work of an engineer can seem like a mystery to those outside the profession. But it's simple, really. Here's how engineers make all these amazing things happen.

First, they see an opportunity to improve people's quality of life. Then, they gather information, brainstorm ideas with a team, and study possible solutions. They choose the best options depending on cost, ease of use, and other factors. They communicate the solution through writing and speaking to groups. Finally, they work with creative, interesting people in other professions to oversee the building or implementation of their idea.

Engineering is sowing the seeds and then helping the plants bear fruit. Engineering is the way dreams become reality.

"Have you ever thought of something that could be better? Do you enjoy working with a variety of people in a positive working environment?" asks **Yee Cho**.

"Do you have a knack for weighing options? Or explaining ideas to others?" **Sunnie House** wonders.

"Do you like to feel the satisfaction of a job well done?" **Dorota Haman** asks you.

What It's Like to be an Engineer

Engineering is considered a profession, just like medicine, accounting, law, or teaching. Engineers have an important obligation to conduct their work in a professional manner. After all, the safety, health, and welfare of millions of people are in an engineer's hands!

Not surprisingly, engineers need an engineering education and sometimes need an advanced degree to learn the intricacies of their specialty. But many engineers get high-paying jobs with just a Bachelor's degree. After practicing engineering for a period of time, many engineers elect to become licensed professional engineers, or "P.E.s."

Licensure is the hallmark of a professional, whether it's a P.E., an M.D. (Medical Doctor), a C.P.A. (Certified Public Accountant), or a J.D. (Juris Doctorate in the profession of law). Professionals have prestige!

Professionals also have flexibility to choose the working environment that best appeals to them. For instance, women engineers can work in the field, in the laboratory, in a classroom, in an office, and even in a boardroom!

"It's having a vision of what can be done, having a desire to realize that vision, and not being tied to how things occur traditionally," say **Edith Martin**, who, among other things, has worked with global technology leaders to shape our national and international telecommunications systems.

Says **Sonya Summerour Clemmons**, "With a Ph.D. in Bioengineering and an M.B.A., I have an amazing variety of career options to choose from, and I have been in a position to create my own non-traditional career niche. My degrees have also provided a level of prestige, financial security, and independence that enable me to live a more fulfilling lifestyle.

Joan Woodard says, "I think there are very few areas of study that can be as beneficial as an engineering degree. Where else but engineering can I take a concept and turn it into reality with such enormous positive impact on the lives of millions of people?"

Why Engineering? Because It's Fun!

An engineering career can offer huge rewards, not the least of which is that engineering is fun! Why? Because of the satisfaction you get from knowing that the world is a better place because of your work. Because of the creative energy that comes from the tremendous variety of work. And because of the personal stimulation that comes from working with interesting people.

Here's what women from all walks of life have to say about the true pleasure they've had being women engineers:

"Engineering is a really fun job. I wouldn't change what I do for a minute," says **Judy Nitsch**.

"I find engineers to be well read, and they have many interesting hobbies. Engineers are exciting folks to be around and to be challenged by," says **Terri Helmlinger**.

"Working in the NASA robot engineering lab is as fun as flying in space. It's true! Where else can you take a drawing on a piece of paper and build it into something that is a walking, talking machine?"

—Nancy Currie

"I wake up in the morning and love my job," says **Decie Autin**. "I'm still energized after 25 years. I come to work because I think it's fun. Engineering's a good career choice and well worth the four years in college and sweating through exams."

Janice Peterson agrees. "The women I know in engineering are successful, bright, and fun to know."

Healthy Bodies, Creative Minds

Biomedical engineering is hot. It's new. It's the future. But what is it?

Biomedical engineering is, simply, engineering applied to the medical field. It covers all parts of the body: the skeleton, the brain and nervous system, muscles and tissue, and the heart and lungs.

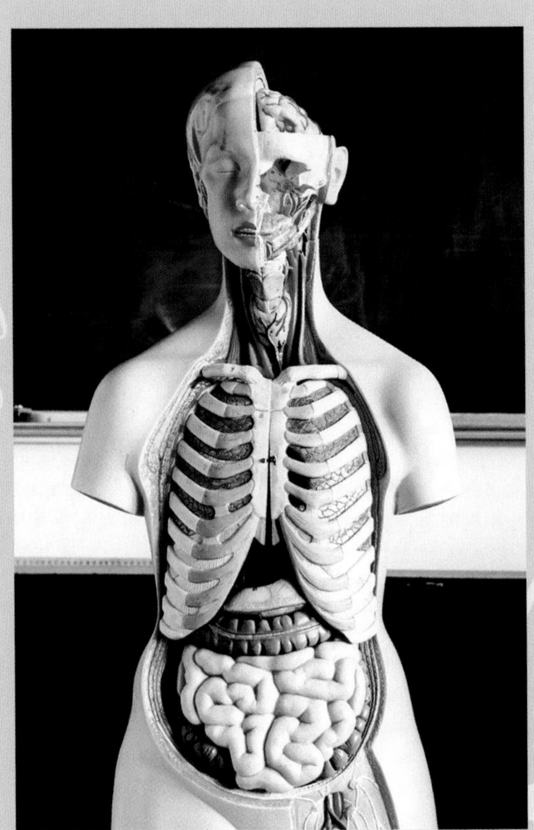

Brain Scan Comparison.

Think about the importance of artificial limbs. Pacemakers. Replacement body organs. Devices that help monitor a person's health.

Engineering has always been a "people serving" profession. But biomedical engineers get close—really close—to people. While doctors treat patients one by one, biomedical engineers treat millions through their clever inventions and creative work. It is, indeed, a tremendously rewarding career.

Above left: A laser-scanned micrograph of cancer cells from a human breast (yellow and bright red). Above right: Mammogram of a healthy breast (right) and a cancerous breast showing a tumor (left).

Brain Power

Talking to friends, smelling flowers, hearing music, tasting chocolate, watching T.V.: we wouldn't experience being alive without a brain.

Though the adult human brain weighs only about three pounds, it contains about 100 billion—that's 100,000,000,000—nerve cells called neurons. The neurons send messages to one another through chemicals called neurotransmitters.

Every time you learn something, new connections are made between brain cells. The more connections made, the greater your intelligence. So keep learning!

Until the 1970s, most brain research was conducted on dead people. Information from dead brains has some serious limitations, to say the least! Today, new tools enable study of the living brain. A whole new realm of engineering has just begun.

Early Brain Wave Measurer

In the brain, one neuron sends an electrically charged particle or ion to another to create a tiny message. Neurons in the human brain are estimated to generate approximately 25 watts of power—nearly as much as a light bulb!

Some people call the electrical activity "brain waves." In the early 1950s, electrical engineer **THELMA ESTRIN** (b.1924) wanted to record brain waves. She thought she could associate wave patterns with different behaviors.

This was not easy in the early days of computers. Thelma's challenge was to convert the analog electrical brain waves obtained through electroencephalography (EEG) to digital form. Then, a computer could more accurately analyze electromagnetic fields on the surface of the human head.

She did it! Thelma's "converter" was one of the first analog-to-digital inventions. "Computers were so new," says Thelma, "most researchers didn't see their relevance to our work. They weren't interested in developing uses for them."

Today we know differently. Thanks to pioneering engineers like Thelma, EEGs, Magnetic Resonance Imaging (MRIs), and Computed Tomography (CT scans) make looking inside the body possible.

Thelma Estrin pioneered the development of computers in medical applications. She applied her knowledge of how the brain works to research on epilepsy and blindness. Thelma's career had lasting influence on her family as well. She raised three daughters, two of whom are remarkable engineers in their own right. (Read about Judy Estrin and Deborah Estrin on page 125.)

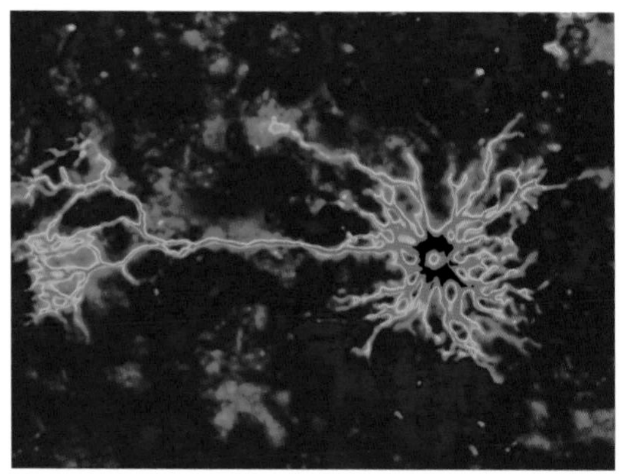

A human neuron (with black nucleus) showing a long dendrite carrying signals to the neuron (in orange), and shorter axons that carry signals away from the neuron to other neurons, muscles, or glands.

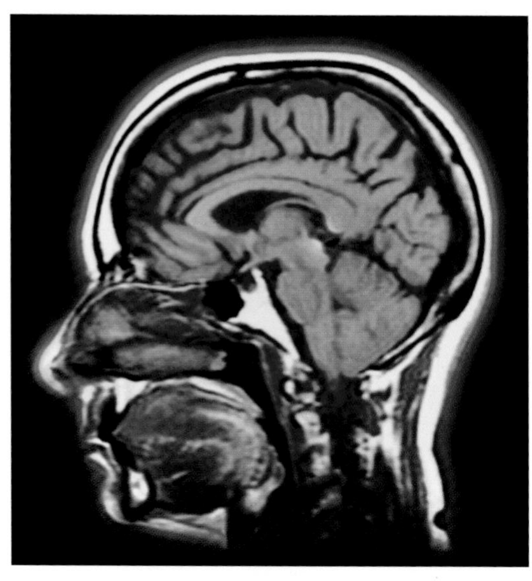

Magnetic resonance image (MRI) of woman's head, side view.

Using Her Head to Save Yours

Traumatic brain injuries caused by car collisions, sports accidents, child abuse, and falls debilitate or kill thousands of people every year. But many safety standards are based on 30-year-old research. Time for a change? Biomedical engineer **SUSAN S. MARGULIES** (b.1960) says "yes!"

Susan uses computer models, children, animals, and instrumented dolls to try and understand how certain head injuries occur in children.

Before Susan's research, people believed that head injuries were caused by large pressure changes in the brain, which led to brain damage. Susan's research turned that theory on its ear.

Susan is studying rapid start and stop movements and their effects on a person's brain. She proved that the part of the brain that is most distorted is the same part of the brain that sustained the most injury. This sparked a whole new round of research to discover how trauma affects blood flow changes, nerve functioning, swelling, and long-term brain cell viability.

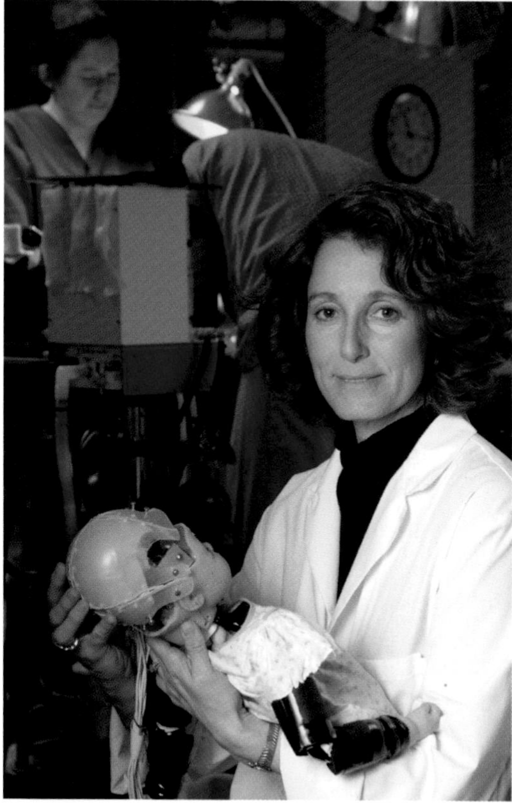

Susan Margulies' group is doing research on young children whose brain tissue and skull stiffness is very different than that of adults. "Our results show it's high time we improve safety standards for helmets, car safety features, and sports and playground equipment," says Susan.

Strong Bones

What would humans be without a skeleton . . . or a funny bone? They sure couldn't walk—or even stand upright. Bones support and protect internal organs, brace muscle movement, and produce blood cells in bone marrow.

Connective tissues bind bones together in a framework, help support organs, and protect against disease. Connective tissue includes muscle that is bound to bone by non-flexible tissue called tendons or elastic tissue called ligaments. Cartilage, a tough, flexible connective tissue strengthened by collagen fibers, is found throughout the body.

Bones grow throughout your childhood and into your twenties. Amazingly, you can increase the density of your bones at any point of your life. Exercise helps!

Our Amazing Bodies

- Infants have at least 300 bones. As children grow, the bones fuse together to form 206 bones in the adult body.
- There are over 600 muscles and 230 joints in the body.
- Our bodies contain more than a gallon of blood.
- The heart is the strongest muscle in the body. It has to be to pump that blood more than 1,000 times per day!

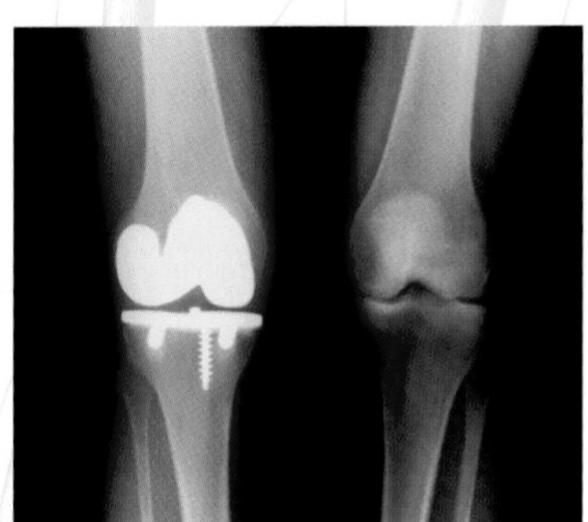

ELECTRIFYING!

In the 1970s a research frenzy was on. What made bones grow stronger? It was known that bone cells grew more densely in areas under mechanical pressure. The question was why?

As a possible explanation, electrical engineer and bioengineer **GLORIA BROOKS REINISH** (b.1925) wanted to determine whether bone is piezoelectric—that is, when stressed, does bone become electrically charged? The tricky part was creating an environment that simulated conditions inside the human body—including fluids—to test "wet" bone.

Gloria designed an elaborate experiment that showed that wet bone *does* have piezoelectric properties. Her important findings served as a basis for further orthopedic studies using electrical currents to stimulate bone growth for speeding up the treatment of fractures, osteoporosis, and post-surgical bone repair.

GLORIA REINISH, A PROFESSOR OF ELECTRICAL ENGINEERING AT FAIRLEIGH DICKINSON UNIVERSITY, TEACHES ONLINE AND CLASSROOM ELECTRICAL ENGINEERING COURSES TO STUDENTS FROM AROUND THE WORLD. SHE ALSO RAISED TWO DAUGHTERS AND A SON, ALL OF WHOM BECAME ENGINEERS.

REPLACEMENT PARTS

Every year, hundreds of thousands of artificial joints—mainly hip and knee joints—are surgically implanted into injured athletes, victims of accidents, and people with mobility problems. These implants help restore normal function in the joint and relieve pain.

Ceramic engineer and metallurgist **ANNA C. FRAKER** (b.1935) investigated many surgical implant materials to find materials compatible with the chemistry of the human body. She wanted to find surgical implants that wouldn't corrode over time. They also had to be strong enough to withstand the pressure and stress of movement and normal body function.

Anna's research resulted in the development of standards for implant materials. Her work has enabled millions of people to lead normal active lives. For that, she is surely a hero.

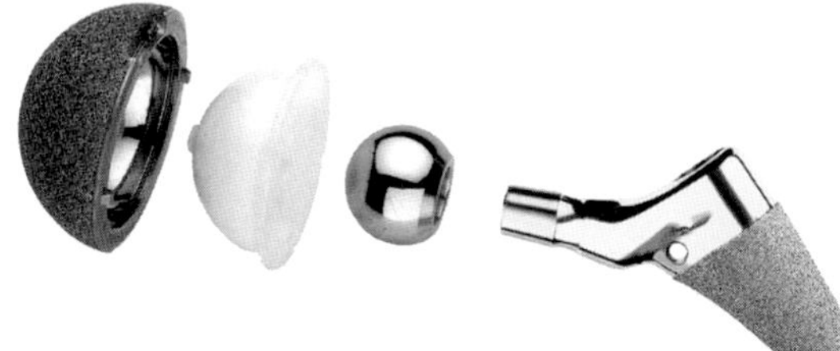

THROUGHOUT HER CAREER, ANNA FRAKER ENJOYED WORKING WITH COLLEAGUES AND STUDENTS FROM VARIOUS FIELDS OF ENGINEERING, SCIENCE, AND MEDICINE TO DEVELOP STANDARDS FOR SURGICAL IMPLANT MATERIALS. THE FEEDBACK SHE RECEIVED FROM IMPLANT RECIPIENTS WAS ALSO CRITICAL TO HER WORK.

THE METAL AND PLASTIC PARTS OF THE ZIMMER® HIP SYSTEM PICTURED HERE (AT LEFT) CAN BE SURGICALLY IMPLANTED TO REPLACE A PAINFUL HIP JOINT. THIS SYSTEM WILL ENABLE COMFORTABLE, SMOOTHER HIP MOVEMENT FOR A LONG TIME.

Fragile Bones

Around ten million Americans have osteoporosis and nearly all are women. Osteoporosis is sometimes called the "silent disease" because it occurs without symptoms. It's characterized by porous, low-mass bones and structural deterioration of bone tissues.

Osteoporosis can lead to increased bone fractures, especially of the hip, spine, and wrist. To avoid weak bones in old age and bone diseases like osteoporosis, it's important to eat a diet rich in calcium and vitamin D, avoid alcohol and cigarettes, and get plenty of weight-bearing exercise.

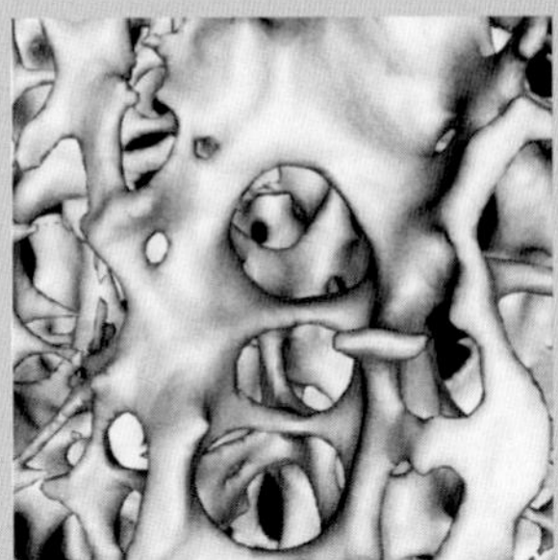

Normal Bone

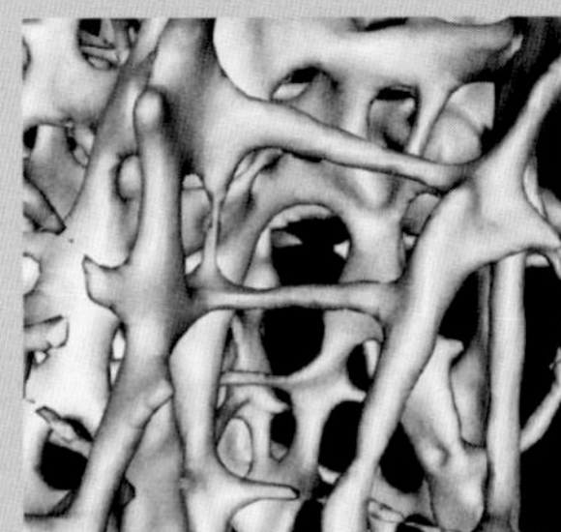

Osteoporotic Bone

Teen Bones

Bones are an area of particular interest to biomechanical engineer **MARJOLEIN C. H. VAN DER MEULEN** (b.1965). She's studying how forces from activities and body weight impact bone growth and development.

Growing bodies gain half of their bone mass during their teen years. Marjolein wondered if boys and girls of all ethnic backgrounds experienced the same gain, and what factors might influence this gain.

Through clinical data combined with computer and animal modeling of the thigh bone called the "femur"—the bone that bears the most weight in the body— Marjolein learned something interesting. When you compare two teenagers of roughly equal height, the one who carries more weight due to a larger body mass will have more bone mass. Their thinner friend will have less bone mass. This is true for girls and boys across different ethnicities.

Increased bone formation during critical periods of growth can help offset adult losses. "If teens can build up more bone mass as their bones form, they'll have more to protect them from osteoporosis when they age," Marjolein concludes.

Marjolein van der Meulen's passion for engineering started in high school when she saw a TV show about paraplegics who were trying to walk again. In college she studied mechanical loading in bone growth—how mechanical forces impact bone development.

"Nothing in life is to be feared.

It is only to be understood."

Marie Curie

Two Time Nobel Prize-Winning Physicist and Professor (1867–1934)

Bodybuilding: Cell by Cell

Spinal cord injuries, bone and cartilage damage, and related diseases are difficult to treat using traditional medical procedures. In the near future, doctors will be using stem cell therapy instead of drugs and surgery to heal many diseases and injuries.

Stem cells, found in bone marrow or blood, are cells that have not developed enough to serve a specific purpose. They have not yet become tissue, muscle, or bone cells. Biomedical engineers are now discovering how stem cells can be grown into different kinds of cells to repair tissue and bone injuries or diseases.

Biomedical engineer **TREENA LIVINGSTON ARINZEH** (b.1970) is a rising star in the field of tissue engineering, a researcher, an assistant professor, and the mother of two children. She's developing stem cell biomaterials that aid in the regeneration and repair of bones, cartilage, and ligaments.

Stem cells alone aren't a miracle cure. If you add stem cells to injured tissue, nothing happens. The stem cells need to sit on a "scaffold" of biomaterial that will help them grow.

Different kinds of biomaterials seeded with stem cells cause different kinds of tissue to grow. For example, Treena has found that calcium and phosphorus can be made into a porous scaffold resembling a hard sponge. When stem cells sitting on this sponge are injected into lab animals, new bone cells form within four to six weeks!

Treena and her colleagues are also working on an amazing biomaterial architecture that looks like meshed fabric similar to cheesecloth. When stem cells are placed in this material and left to grow in lab animals, cartilage can regenerate, "an unparalleled feat still to be accomplished in humans," Treena notes. But, she's working on it!

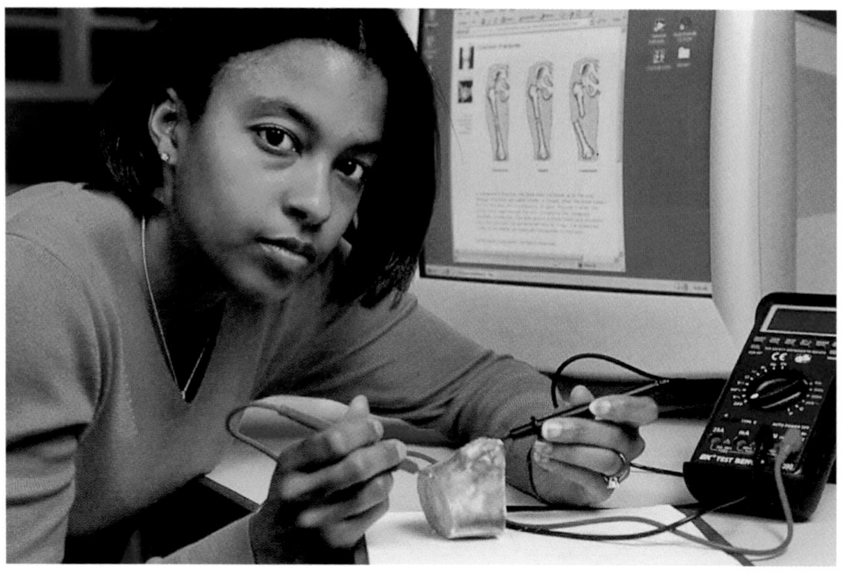

"It's an exciting time to be in tissue engineering," says Treena Arinzeh. "New technology, tools, and computations that weren't available 10 years ago have opened new paths to research. I'm fortunate to have had the encouragement of teachers and friends to do the remarkable work that I love."

Her newest challenge is to develop scaffolds to carry stem cells for spinal cord repair. Ultimately, Treena hopes her research will aid in helping paralyzed patients restore movement to their limbs.

At present, her work is being tested *in vitro* (in a lab setting) and *in vivo* (in live bodies) of small animals. "Making things work in the lab is easy," says Treena. "But when it comes to animals and humans, it's a whole new ball game."

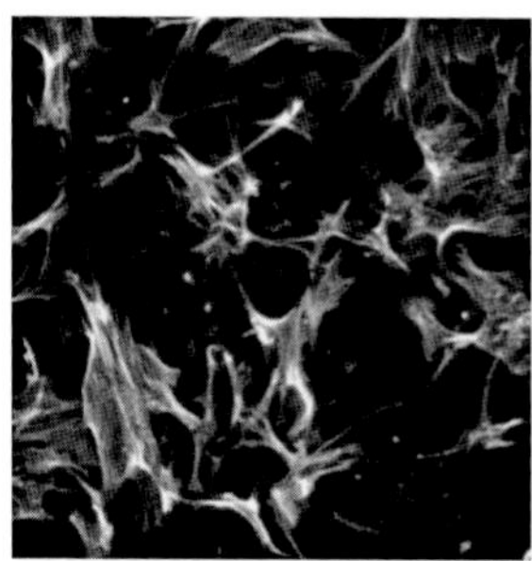

These stem cells turn into bone cells after four days of culture on biomaterial.

Flex Your Muscles!

Muscle is like an engine. It burns calories and powers movement. Without muscles, there would be no speech or singing, writing, walking, dancing, or sports.

Muscles are attached to our bones to enable the body to move. Did you know that for every muscle that causes motion, there's an opposing muscle that causes the opposite motion? Muscles also cause our body organs—such as our lungs, stomachs, and bladders—to contract as needed to do their jobs.

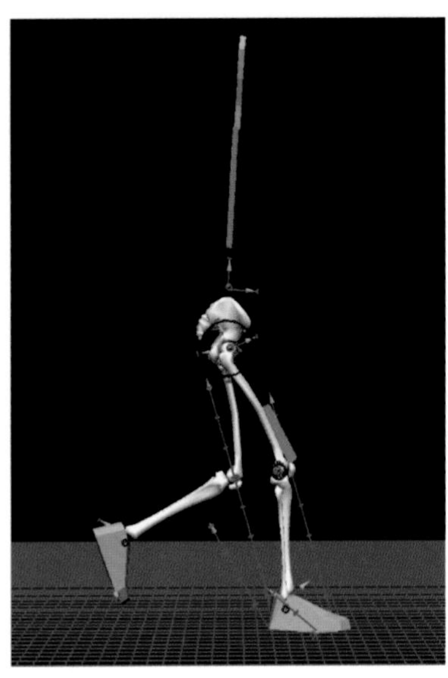

Helping Kids Walk

"I loved sports as a kid. Especially gymnastics," says biomedical engineer **ALLISON ARNOLD** (b.1968). "I was always amazed by what my muscles could do. I think that's a big part of why I'm now researching muscles: how they contribute to motion and how they can be rehabilitated."

Allison helps orthopedic surgeons (doctors who correct problems with our skeletons, muscles, and joints) treat kids with walking disabilities caused by diseases like Cerebral Palsy (CP). CP causes the brain to activate muscles incorrectly, so people who have the disease tend to trip easily or fall down.

To help doctors plan treatments, Allison is using a computer to model how patients walk. She digitally recreates a patient's motions and skeletal geometry, then solves equations for the muscle forces, which are hard to measure directly.

"While this work is still at the development stage, we're hoping the models will provide guidelines to surgeons about which muscles should be weakened or which tendons should be transferred to a different place on the bone to help patients walk better."

"I'm amazed by how little is known about muscles and how they are coordinated to produce body movement. There's so much to discover," says Allison Arnold. She works with mechanical and software engineers—along with doctors, physical therapists and patients—to study ways of improving muscle function in kids with walking disorders. "From the computer models we're building, we're hoping to eliminate trial and error by predicting how successful various surgeries will be."

Pain in the Neck

Feel as if you have the weight of the world on your shoulders? That's because you do! Humans have incredibly heavy heads propped up on thin necks. This predisposes humans to neck injuries, especially whiplash from auto accidents.

Biomedical engineer **ANITA VASAVADA** (b.1967) specializes in 3-D computer modeling to simulate the movements of neck muscles and ligaments under trauma. Studies have found that fifty percent more women suffer whiplash than men. Anita plans on using computer models to determine the causes of the higher incidence of whiplash injury in women.

Anita works with engineers who videotape human volunteers in actual car crashes (at very slow speeds, of course). She then inputs data about the subjects' head and neck movements into a computer model to determine what happens to the muscles. "I've learned that muscles over-stretch during the impact, which is a possible cause of neck injury," says Anita.

She's confident her research will help in developing new treatments and physical therapies for neck injuries. "I hope that someday automobile manufacturers will be able to design more supportive seats and neck rests that minimize or prevent whiplash altogether."

If the geometry of a woman's neck is different than a man's, then a whole new computer model must be developed to study it. Anita Vasavada (below right) is hot on the trail. Anita's research will aid in developing new treatments and physical therapies for neck injuries in women and men.

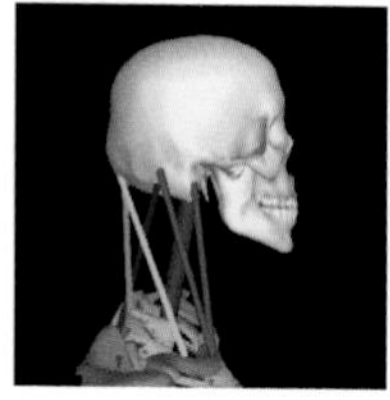
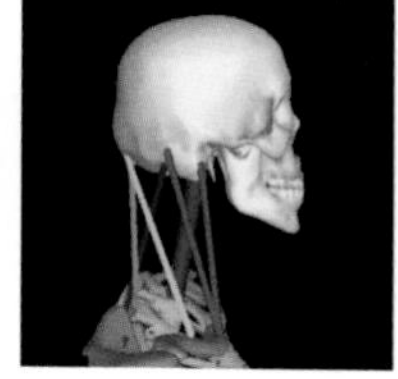
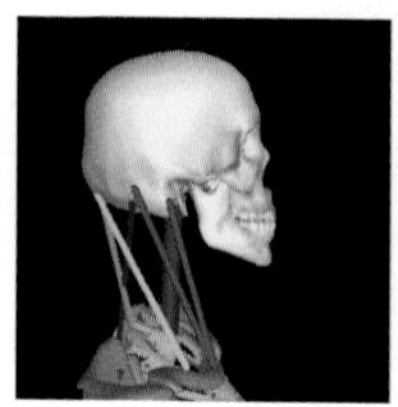
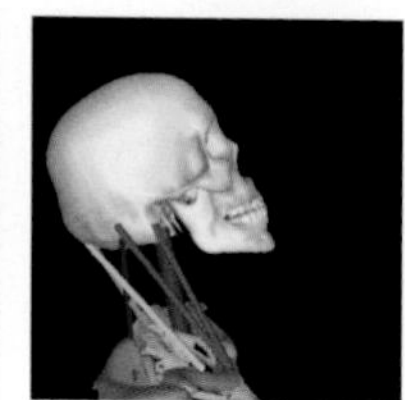
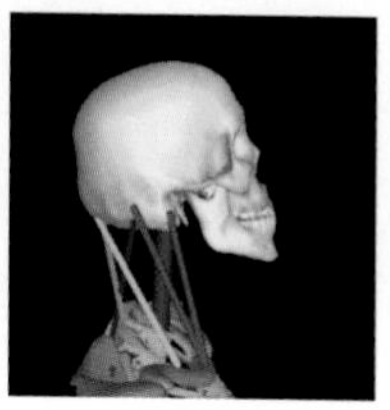
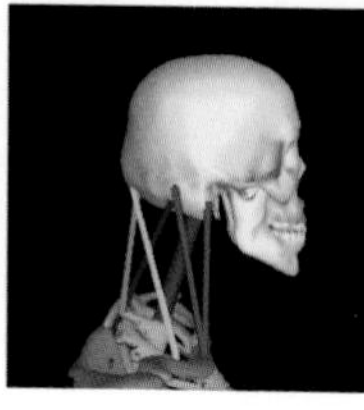
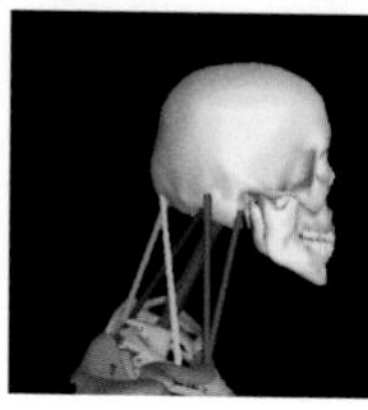
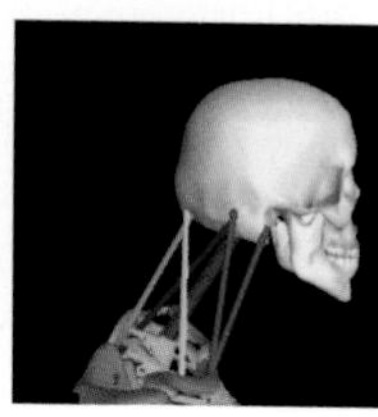

Big Heart

About the size of a fist, the heart is composed of many different kinds of cells that work together to circulate blood. The blood carries oxygen from the lungs and other nutrients throughout the body.

A healthy heart beats constantly, about 100,000 times every 24 hours! It's not uncommon for the heart to skip a beat once in a while, but a pattern of arrhythmia—irregular heartbeats—may lead to heart disease, strokes, or heart attacks.

The American Heart Association estimates that as many as 2.2 million Americans live with arterial fibrillation, one of the most common forms of arrhythmia. Cardiovascular (heart) disease causes more than 500,000 deaths in the U.S. every year. Wow! Heart-healthy engineering has the potential to make a huge difference!

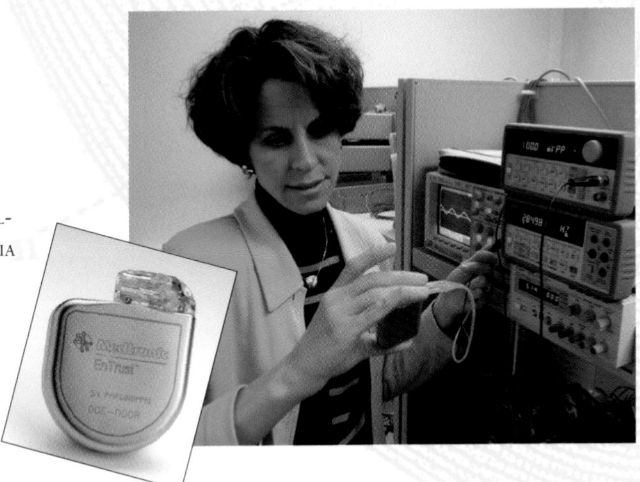

Kristina Ropella has researched ways to improve implant devices, like cardiac defibrillators (inset), to help people with arrhythmia achieve more normal heartbeats. Here she's holding an implantable defibrillator and using electronic instruments to measure its function. The Medtronic EnTrust™ implantable cardioverter-defibrillator (ICD) courtesy of Medtronic, Inc.

Listen to the Heartbeat

Biomedical engineer **KRISTINA M. ROPELLA** (b.1963) has been researching cardiac arrhythmias, or irregular heartbeats, for years. One of her favorite projects involved defibrillators and pacemakers, which are implanted in a human's heart to treat arrhythmias.

The question was how to make them better. Smart devices in the implants must constantly monitor—or listen to—the heart's activity. If the heartbeats become abnormal, the patient may need serious help!

Given the huge variability of the human heart, Kris computerized the electrical data she collected from heart implants and analyzed it to learn more about the signal patterns. Kris was most surprised to discover that her computer algorithms for detecting abnormal heart rhythms work over 95 percent of the time. That's amazing, considering the billions of people out there moving to their own beat.

Kris shared her results with a medical device company that makes defibrillators. The company was then able to build better defibrillators that detect potential life-threatening changes in arrhythmias.

Gordana Vunjak-Novakovic (left) and Milica Radisic (right) are analyzing the behavior of cardiac (heart) tissue they've grown. The tissue samples are connected to a pacemaker to measure the voltage, amplitude, and frequency of contractions. Someday, heart tissue patches may be used to repair the damage caused by heart attacks.

Above: The heart tissue in columns one and two was grown in the lab. The first column did not receive electrical stimulation, while the second column did. The third column shows "real" (native) heart tissue. Notice how close the electrically stimulated tissue is to native heart tissue.

At left: This heart tissue sample shows contractile protein chains (stained green) that help the heart beat. The sample was grown from individual cells in the lab with the aid of electrical stimulation, and closely resembles the structure and beating of a "real" heart.

How to Mend a Broken Heart

It's no easy matter to understand how to mend a broken heart, but tissue engineers **GORDANA VUNJAK-NOVAKOVIC** (b.1948) and **MILICA RADISIC** (b.1976) are working on it.

Gordana and Milica have successfully grown heart tissue that actually beats! They seeded cells on a scaffold support made mainly from collagen—a tough elastic substance found in both animal and human bodies. The tissue grew to about the size of a fingernail after only eight days in the lab.

When the tissue is attached to electrodes from a pacemaker, it mimics the beating of a heart! Someday, heart tissue patches could be used to repair damage from heart attacks or heart malformations in newborns.

Gordana is also studying how tissue cells are affected by gravity. She's even made special cell culture equipment that NASA has taken into space. Mending broken hearts all across the galaxy may not be science fiction after all!

Biomedical engineering and tissue engineering are two of the fastest growing areas of engineering. In fact, nearly 40 percent of engineering bachelor's degrees awarded to women in 2000 were in biomedical engineering.

Breathe Deeply

Humans usually breathe in and out 15 to 25 times per minute without giving it a single thought. Our lungs serve one purpose: they take in oxygen, a gas that we need, and get rid of another gas, carbon dioxide, that we don't need. To do that they carry the venous blood that's saturated with carbon dioxide away from the heart, and return oxygenated blood. This is called pulmonary circulation.

Lamb Lungs

"When people have lung diseases or a hole in their heart, pulmonary circulation can be in trouble," says biomedical engineer **CAROL LUCAS** (b.1940). She researched pulmonary circulation in newborns and young children to help doctors determine if surgery could help.

But here's the drawback: "Sometimes surgery increases lung pressure and causes fluid build-up in the lungs. That's not good," Carol explains.

To figure out normal lung pressure in kids, and how lung pressure changes during growth, Carol decided to study lambs. "Baby lambs and human infants have similar birth weight. In fact, growing lambs are often used as models to study lung development in young children."

Carol built algorithms, took velocity measurements, and used ultrasound to determine normal pulmonary pressure in lambs. Her studies resulted in suggested parameters indicating whether surgery can be beneficial or harmful for a patient with lung problems.

"Healthy lung pressure is much, much lower than blood pressure," says Carol Lucas, who studied lamb lungs to establish healthy levels of pulmonary pressure in young children and infants. Carol is shown here with models of lamb lungs. "Math and engineering skills are essential to my research. I always tell students there is no such thing as too much math in their course load. Math skills are vitally important!"

TAKING THE WHEEZE OUT OF FREEZE

Many people find that, along with scarves and mittens, freezing weather brings out cold weather asthma attacks.

Asthma experts used to think that cold air triggered asthma attacks in one of two ways: a cold breath caused fluid to evaporate from the lungs, leaving behind a saltier solution that irritated the lungs; or cold air caused airway blood vessels to shrink, which brought on attacks.

Biomedical engineer **JANIE M. FOUKE'S** (b.1951) groundbreaking studies proved both theories wrong and left the asthma community breathless with amazement. Janie found that when you inhale cold air, a rush of blood pours into the lungs to warm them up. The rush causes the blood vessels to heat up and swell, which in turn prompts an asthma attack.

Janie's findings changed the course of drug treatment for asthma patients and resulted in a new wave of research on blood vessels and their effect on asthma.

JANIE FOUKE KNOWS ONE TRIGGER FOR AN ASTHMA ATTACK IS COLD AIR. WHAT BETTER PLACE TO STUDY THE EFFECTS OF COLD AIR ON BREATHING THAN THE NORTH POLE! ALL OF JANIE'S EXPERIMENTS ARE DESIGNED TO IMPROVE TREATMENTS FOR LUNG DISEASE.

A Breathtaking Disease

While breathing is something most people take for granted, many people don't. They or their children suffer from asthma.

Asthma is the most common chronic disease of childhood. It affects nearly five million children and more than 12 million adults in the U.S. Dust, dander, mold, and pollution enflame the airways, making it difficult to breath.

Cancer: the Cellular Scourge

Have you ever known a person who had cancer? Mother? Grandmother? Aunt? One woman out of seven is likely to get breast cancer in her lifetime. In fact, breast cancer is the most common cause of cancer deaths for women in the U.S. But there are many other types of cancer that are equally dangerous: skin cancer, colon cancer, and lung cancer, to name a few.

Cancer is a disease where abnormal cells divide and grow unchecked. The cancerous cells grow into abnormal tissue, called tumors. Cancer is the leading cause of death after heart disease. It kills more than half a million people in the U.S. every year.

Mammography has helped millions of women discover breast cancer at an early stage. The problem is that tumors must be about a centimeter in length and several years old before a mammogram can spot them.

If cancer cells can be sighted earlier, the cancer can be treated faster and easier. People will have a better chance of surviving cancer as well.

Background: Kidney cancer cells.

Shine a Searchlight on Cancer

With "molecular imaging," we can see what's happening at the cellular and molecular level in a living organism. Molecular imaging of cancer fascinates biomedical engineer **EVA SEVICK-MURACA** (b.1961). Eva and her graduate students discovered a way to specifically image cancer using near-infrared light.

To find the cancer cells, Eva's group uses fluorescent imaging agents connected to proteins. When these are injected into the body, the agents hone in on cancer. Upon shining red laser diode light (similar to that in the supermarket checkout) onto tissue surfaces, the fluorescent imaging agents within the body emit a different color light that can be collected to image cancerous tissues using a mathematical model.

The fluorescence imaging method may be more sensitive than other approaches and can put the "spotlight" on cancer cells that are just starting to grow.

"In a few years," says Eva, "it may be possible to find breast cancer at a much earlier stage, before the disease becomes more serious. Breast cancer could become more manageable, making it easier for doctors to achieve a cure."

Eva Sevick-Muraca and Kildong Hwang discuss the delivery and collection of near-infrared light in a model of a breast. They're using fiber optics to demonstrate the abilities of this imaging technique to the U.S. Food and Drug Administration.

NANOPROBES TO THE RESCUE

Electrical and computing engineer **MIHRI OZKAN** (b.1966) is doing research with teeny nanoprobes to detect diseases such as cancer. "Working at nanoscale level means I can't see what I'm doing with my eyes, so my research is full of exciting surprises," says Mihri. She uses electron microscopes to view her work.

Ultimately, these multi-functional probes will enable doctors to watch as the probes travel through the body, target cancer cells, and deliver drug therapy. "Any development we can add to cancer detection and therapy will have a global impact on human life," says Mihri.

Mihri grew up in Turkey, where more girls go into engineering than in the U.S. "When I made my career choice, it wasn't a big deal and everyone was very supportive." Mihri finds great satisfaction in her research, and also loves spending time with her two sons.

IN COLLABORATION WITH THE SCRIPPS RESEARCH INSTITUTE, MIHRI OZKAN AND HER TEAM ARE GETTING READY TO PUT NANO-PROBES INTO VIRUS CLUSTERS AND INJECT THEM INTO LAB RATS. "WE TAG THE PROBES WITH MINUTE PARTICLES OF GOLD OR FLUORESCENT NANOPARTICLES, SO THEIR COLOR CAN BE SEEN AS THEY TRAVEL AROUND INSIDE THE RAT."

CANCER BURN OUT

Imagine a cure for cancer within the next few years! It's possible through nanotechnology and brilliant research by biomedical engineers such as **JENNIFER WEST** (b.1970). "I can't believe how smoothly the cancer research has gone," Jennifer says. "Everything seems to be falling into place with very few problems. It's not like any research I've ever done before."

Jennifer discovered a unique form of cancer treatment that has been 100 percent successful in lab animals. What's Jennifer's silver bullet? Nanoshells, teeny-tiny particles with optical (light) properties that can burn out tumors.

Nanoshells are injected into the blood stream where they can easily break through holes in the blood vessels surrounding the cancer. The nanoshells are attracted to cancerous cells by special protein markers.

When a fiber optic laser light is shined on the area, the nanoshells go into action and burn out the cancerous tissue. What's great about that is the light is transparent to healthy tissues, but it shows up in cancer cells.

While the first human applications will be for inoperable cancer such as tumors in the brain or heart, Jennifer's nanoshell therapy might ultimately cure cancers of many varieties.

JENNIFER WEST DIDN'T WANT TO BECOME A DOCTOR, BUT WANTED TO WORK IN HEALTH SCIENCES WHERE SHE COULD APPLY MATH AND ENGINEERING SKILLS. BIOMEDICAL ENGINEERING WAS HER CALLING, AND IT MAY INDEED PLACE HER IN THE BIOMEDICAL ENGINEERING HALL OF FAME.

The nanometer scale is 10^9 smaller than one meter. Does anyone want to do the math?

Miracle Cures

The next wave of biomedical technology will radically alter healthcare as we know it today. Biomedical engineers talk about "convergent technologies," where engineering, chemistry, biology, and technology are all combined to produce new drugs and medical products.

Tens of billions of dollars are being spent every year on these "miracle cures" that will hopefully be the remedies for AIDS, cancer, heart disease, stroke, infectious disease, neurological disorders, and mental illness.

"Engineering is all about the invention of devices and processes. Any young engineer with imagination can make a huge contribution. Women are really needed because they think differently than men and they're bringing a whole new vision of creativity that will have far-reaching impact on our world."

—Corinne Lengsfeld

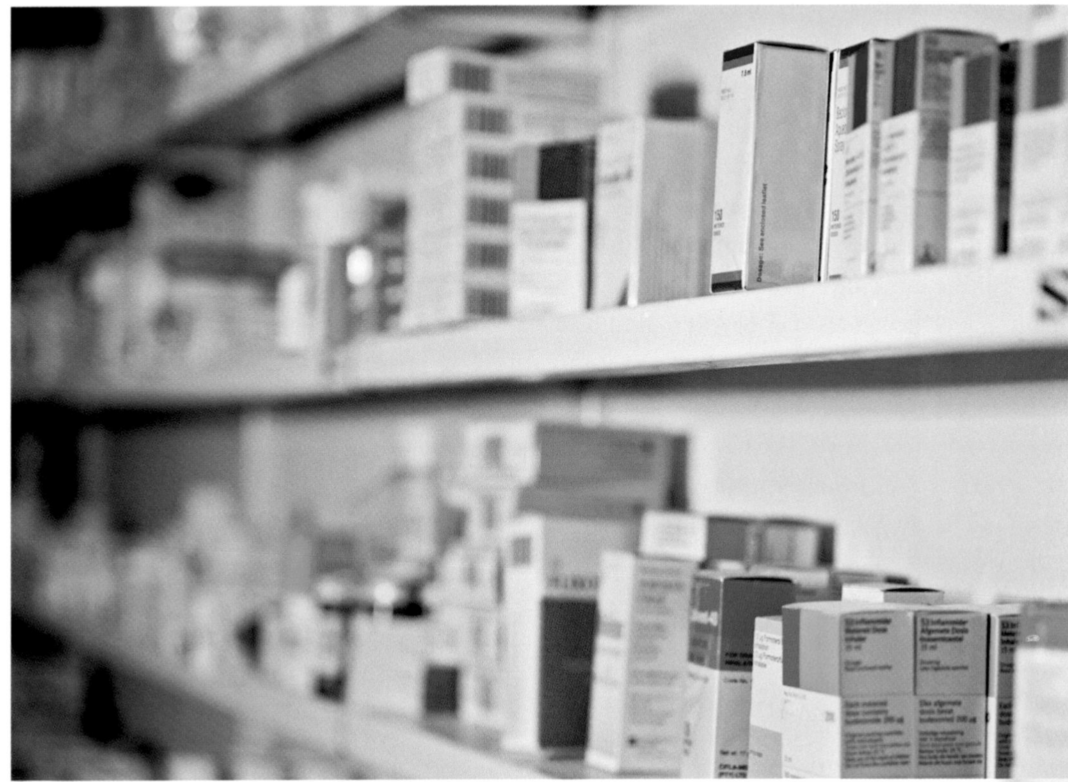

After Sonya Summerour Clemmons got her doctorate in bioengineering, she worked with biotech start-ups to bring cutting-edge products to market. Now, with an MBA, she's in a perfect position to negotiate business between scientists and corporate executives. Sonya and her husband are also raising a three-year-old daughter.

Special Delivery

Biomedical engineer **SONYA SUMMEROUR CLEMMONS** (b.1971) sees a tremendous future for new ways to deliver therapeutic drugs and other agents to the human body. She works at a biotech company that develops polymers—chains of molecules linked together—to help deliver therapeutic drugs and other biological compounds to the body.

For example, when patients have heart surgery to clean and open arteries, a permanent metal stent (a mesh-like tube) is often inserted to hold the artery open. Now, polymers infused with drugs can be placed on the metal stents to help wounds heal better and faster.

In the near future, there won't be any drug-eluding metal stents at all, only polymers that break down into amino acids and blend into the body without leaving a trace of foreign material!

Sonya notes that, "Engineering has been the backbone of my career. It's a great foundation, but in today's biotech landscape, it's wise to combine an engineering background with science and business degrees. That way your career options are plentiful."

Sonya also writes an online column that offers advice to young women in the fields of science and engineering, and inspires people of color.

Designer Genes

Gene therapy is still in the experimental stage. But it looks like a promising way to cure diseases like cancer, cystic fibrosis, and other genetic diseases—by correcting the defective genes responsible for the disease. But how do you get the normal genes to replace the abnormal disease-causing genes?

By a "carrier molecule" like a virus! "Viruses are the best delivery tool there is for getting genetic material into the body," says mechanical engineer **CORINNE LENGSFELD** (b.1971). "But even man-made viruses can reassemble genetic code with disastrous results."

So Corinne is engineering synthetic viruses, tiny cells—1/1,000 of a millimeter in size—that contain water and genetic material inside a shell made of lipids, or fats. She's now testing them in the lab to see if they can do their job. She's included a genetic code for the color green, so if they work, they'll turn the cells green.

"My mother taught me that every person is different. What's blown me away is that every cell is different, too. That makes my job in genetic therapy very complex and fascinating."

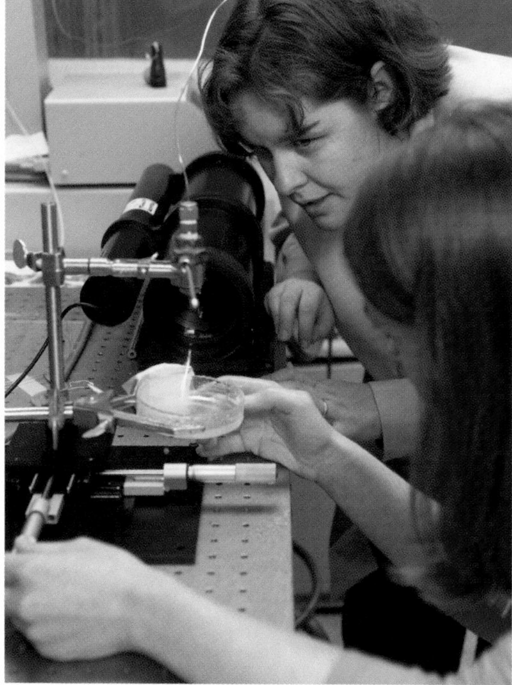

Corinne Lengsfeld got her doctorate in mechanical engineering. Since then, she's worked in oil fields and nuclear power plants, has analyzed the evolution of fish locomotion, studied the transfer of smoke in restaurants, and designed liquid rocket engines. Now, she specializes in biomechanics, drug delivery, and gene therapy. "Engineering is a great degree. You can go anywhere with it."

Sustainable Sustenance

"Everything in moderation," they say. This goes for food, for exercise, and even for politics! Moderation can be thought of as not too much and not too little. Just the right amount.

Food. Fresh, juicy, healthy, sweet—it's one of the delights of being alive. We tend to eat more of what tastes good. But, we should also think more about what we're eating and the amount we're ingesting.

Farm in Lancaster County, Pennsylvania.

When agricultural, food, and biological engineers think about moderation, they also think about balance. The world's growing population demands more food and energy. However, there are limited supplies of natural resources.

This is where sustainability comes in. Sustainability is meeting the needs of the present without undermining the ability of future generations to meet their own needs.

Agricultural engineers devise practical improvements for sustainable irrigation systems, equipment and machinery, crop planting and harvesting, livestock management, and food production and processing. Maintaining sustainable balance with the environment is the key.

VIEW FROM ABOVE OF A PEA-SORTING MACHINE IN A FOOD PROCESSING PLANT.

Yummy!

We often take for granted the abundant fresh, safe, easy food in our grocery stores. There are so many choices that few consumers stop to think about how complex it is to deliver these products to the grocery store shelves.

A good recipe isn't enough. Raw foodstuffs have to be converted into edible consumer products. The products then have to be preserved, packaged, and distributed. That's just what food engineers do!

To the Chip

In 2002, Frito Lay Corporation decided to pursue a new line of natural and organic potato chip products. Frito Lay had never produced organic products before. To be called "organic," food has to comply with very strict standards.

Chemical engineer **RACHEL L. THOMAS** (b.1973) and her team had to ensure that all United States Department of Agriculture's (USDA) organic regulations and guidelines were followed—"to the chip." Not only did the potatoes need to be grown naturally, but the team had to meet strict oil requirements.

They reinstalled oil tanks for non-hydrogenated frying, and operators had to be carefully trained to follow specific procedures. While existing equipment could be used for other processing, a special sanitation procedure was used so there was no cross-contamination between the natural chips and other products. Natural sea salt was used to season the chips.

The principles of chemical engineering Rachel Thomas learned in school helped her with making potato chips: mass transfer (how many raw ingredients are needed to make a product; in this case, four pounds of potatoes to make one pound of chips), equipment efficiency, and heat transfer (how hot to fry the crispy chips). "To get a better idea of what engineers do, girls should shadow an engineer at work for a day," Rachel recommends.

KUDOS TO YOU

She's worked on pet food, candy, vegetable shortening, and beer, but one of chemical engineer **G. JEAN HOPPERT'S** (b.1949) favorite projects was a tasty granola bar called Kudos®.

At M&M/Mars, Jean's plant specialized in making Milky Way® and 3 Musketeers®. The equipment was perfect for processing thick nougat and caramel substances, but would the same equipment pulverize the fragile rice, crunchy oats, and fluffy coconut into mush? "It's all very well to cook up a delectable, crunchy bar in a test kitchen," says Jean, "but the real challenge is how to mass produce and distribute it through grocery stores across the U.S. and beyond."

Jean and her team made some modifications and developed a special roll-extruding machine that could turn out six-foot slabs of uncut granola bars without denting a grain.

Another challenge was to keep the oats and rice crispy in packaging. Jean and her team figured out a way to include an oxygen barrier in the packaging that kept moisture out and crispness in. She also worked on the logistics to keep the product moving speedily through the plant and get it delivered—fresh-baked—to grocery shelves. From test kitchen to store shelves, Kudos went from an idea to a product within 15 months.

JEAN HOPPERT STANDS AT THE CONTROLS ON THE BAR DELIVERY SYSTEM THAT TRANSFERS CANDY BARS TO THE CARTONS AND NOTES THE NEW CARTON SIZE TO ENSURE CLEARANCE ON THE CONVEYER. "ENGINEERING IS ALL AROUND US IN HOUSES, CARS, FOOD, COMPUTERS, CLOTHING, BUSINESS . . . EVERYWHERE, EVEN IN CANDY BARS. APPLIED ENGINEERING IS GREAT PROBLEM SOLVING AND IT'S FUN!" JEAN IS NOW RETIRED AND A SENIOR MENTOR FOR FIRST, AN INNOVATIVE SCI-TECH ORGANIZATION THAT SPONSORS SCHOOL ROBOTICS COMPETITIONS AND MENTORS TEENS.

Can, Too!

JULIA BRAINERD HALL (1859–1925) was the older sister of Charles Martin Hall, the man credited with developing aluminum. An estimated three billion pounds of aluminum are used every year to make about 100 billion soft drink cans. The Hall's manufacturing process—dissolving aluminum oxide and cryolite, then applying electricity to the mixture—became the basis for a company that later became ALCOA (the Aluminum Company of America).

Charles Hall went on to become a wealthy man, yet Julia never received rightful recompense, nor did Charles credit her when he won a prestigious award for the invention. Like many women of her time, Julia never got the true credit she deserved for her savvy technical and entrepreneurial skills.

Julia was a big picture thinker with an eye toward marketing aluminum from the outset. Her broad approach was reflected in the tasks she undertook. She helped with the research and development, found financial backing, and ran the business.

Julia's meticulous notes were invaluable during the patenting process when Frenchman Paul Heroult tried to prove that he had invented aluminum. Julia's recordings proved that the Halls got there first.

Ah, Chocolate!

One of the world's leading manufacturers of fine chocolates is England's Cadbury, Ltd., now Cadbury Trebor Bassett. That's where chemical engineer **EMMA MCLEOD** (b.1967) works as a senior manager in process development, when she's not on a panel for chocolate tasting!

Emma helps produce new confections, including Cadbury's Fuse candy bar. She was recently charged with setting up a new plant to make Fuse bars. "My team had to decide how big all the equipment needed to be, and how to fit all of it—tanks, pumps, and pipes—into an older building being renovated as a new plant site," says Emma. She also had to ensure that production line software was working correctly, and train a staff of operators to handle the process.

"I've learned a great deal about chocolate, and now my oldest son enjoys concocting new candy recipes, so we've created our own chocolate bars at home. They're good, too," comments Emma.

Emma McLeod checks software and controls to achieve the correct balance and flow of ingredients in the chocolate-making process. Achieving balance and flow at home is also an art. "You have to be really organized to raise children—I have two sons—and have a career. But, you can do it. Cadbury is a flexible, child-friendly company, and that's a big help," says Emma, who also plays the French horn in a local band.

So Goes the Flow

Salsa, catsup, and pasta sauce are made from processing tomatoes, a type of tomato that is firmer and less juicy than salad tomatoes. To be sure product quality and consistency meet industry standards, processing tomatoes are closely monitored on the assembly line. "That's where I come in," says food engineer **KATHRYN L. MCCARTHY** (b.1958).

Kathryn has developed the best use of ultrasound sensors to monitor the correct thickness—viscosity—of tomato paste and sauces while it flows through pipes and extruding machines.

She's also researched the use of magnetic resonance imaging (MRI) sensors to monitor the flow characteristics of ice cream and chocolate. Consistency and flow are not only important to quality, they're also important to maintain production volume and keep labor costs down. Adds Kathryn, "you might be surprised to know that ice cream flows very nicely at minus 12°F."

Kathryn McCarthy (right) and chemical engineering student Shanying Tu (left) are sorting processing tomatoes by varieties, and checking for defects and firmness. Both Kathryn and her husband are faculty at the University of California, Davis. "Our rule is not to work more than 40 hours per week. It's all about setting priorities." Kathryn is also an avid skier and enjoys spending as much time with her two teenagers "as they'll allow."

DANIELLE CARRIER, MOTHER OF THREE DAUGHTERS, FORESEES A FUTURE WHEN PHYTOCHEMICALS—PLANT CHEMICALS—ARE READILY AVAILABLE TO MAKE FOODS HEALTHIER. HER RESEARCH IS AIMED AT DEVELOPING EFFICIENT WAYS TO EXTRACT THE PHYTOCHEMICALS SO THEY MAINTAIN THEIR HIGH QUALITY AND HEALTH BENEFITS.

FOOD AS MEDICINE

The trend in the food production industry toward vitamin-enhanced foods and healthy, natural products is now going a step further. Biological engineer **DANIELLE JULIE CARRIER** (b.1959) is researching the use of "phytochemicals"—the chemicals that naturally occur in plants—to promote human health.

For example, Julie has learned that silymarins from milk thistle and lycopene from watermelon help to prevent fats in LDL—"bad" cholesterol—from oxidizing, or introducing oxygen that can contribute to breaking down healthy cells. This is important in the prevention of heart disease, such as hardening of the arteries where lipids build up plaque on artery walls.

Julie extracts phytochemicals from plant waste material such as grape residue from the wine industry, watermelon crops, and small trees. She then isolates the chemicals and tests them to discover health benefits. "The relationship developing between agriculture and medicine is incredibly exciting," she says.

"A discovery is said to be an accident meeting a prepared mind."

ALBERT VON SZENT-GYORGYI
BIOCHEMIST (1893–1986)

BROWN RICE, PLEASE

"When you buy white rice in the supermarket, the outside husk and bran layer have already been discarded as waste, leaving only the white kernel behind," explains **MARYBETH LIMA** (b.1965).

As a biological and agricultural engineer, Marybeth is researching far better uses for rice waste. "There are cancer-fighting compounds in rice bran as well as protein, oil, and vitamin E."

Marybeth is examining different varieties of rice and parts of the bran layer to determine where the most antioxidants—agents that help repair cells in the body—are found. "Adding rice bran to foods and its extracts to pharmaceuticals not only promotes health, it's also better for the environment because waste is minimized," notes Marybeth.

Eat brown rice next time. It's much healthier because it still has the bran layer on the outside. That's why it's brown!

WHEN MARYBETH LIMA ISN'T RESEARCHING OR TEACHING LAB CLASSES, SHE'S PURSUING A PET PROJECT. "I'M BUILDING SAFE, FUN, AND ACCESSIBLE PLAYGROUNDS THROUGHOUT THE COMMUNITY WITH MY STUDENTS. WE PARTNER WITH SCHOOLS AND PARENTS, GET INPUT FROM THE KIDS ABOUT THEIR NEEDS, AND THEN DIALOGUE WITH LOCAL AGENCIES WHILE WE DESIGN GREAT PLAYGROUNDS THAT MEET STRICT SAFETY CODES," SAYS MARYBETH. HERE SHE'S MEASURING THE SPACE BETWEEN BARS THAT COULD ENTRAP TODDLERS' HEADS. STRICT SAFETY CODES MUST BE FOLLOWED TO GUARD AGAINST THIS HAPPENING.

Safe to Eat? Safe to Drink?

Food-borne bacteria—that can make people very sick or even cause death—ran rampant into the early 1900s. Before the U.S. Pure Food and Drug Act passed in 1906, the food industry was largely unregulated. Vegetables and fruits were often riddled with fungus or pests. Concoctions sold by traveling salesmen were routinely called "patent medicine," but were little more than flavored water.

The passage of a federal law gained force in the early 1900s due to the efforts of the U.S. Department of Agriculture, various associations and manufacturers, and of course, caretakers of the family, the General Federation of Women's Clubs.

Thanks to their efforts, the Food and Drug Administration was created to ensure the foods and drugs we consume are carefully monitored and tested to assure reliable quality and purity.

Background: *Aspergillus*, a family of mold with 185 species, some of which can cause very toxic reactions in humans.

Care, Cleanliness, and Cold

At the turn of the century as the industrial revolution was in full swing, people flocked from the countryside to the cities to work in factories and, they hoped, improve their quality of life.

Perishable food such as milk, eggs, chicken, and fish had to be transported great distances from farm to city. Refrigeration was extremely crude.

Tons and tons of food were sent to the dump because it had spoiled. Even worse, though, were the diseases caused by people eating and drinking rancid food.

MARY ENGLE PENNINGTON (1872–1952) and her employees at the Food Research Laboratory worked tirelessly to understand the exact conditions that caused food to go bad.

Mary was a renowned expert in chemistry, biology, and bacteriology—the study of bacteria—who became an expert on food spoilage and refrigeration. She was one of the first female members of the American Society of Heating, Refrigerating and Air-conditioning Engineers.

Mary's work led to a revolution in the way food was handled and stored in packinghouses, transported in railroad cars, warehoused, and distributed to market. "Care, cleanliness, and cold" was Mary's motto. In the food industry, it still holds true today!

For several years, Mary Pennington rode freight trains back and forth across the country checking temperatures in the cars and analyzing the foodstuffs. Her mission was to determine the optimal conditions so that food wouldn't spoil in transit. She even had a private car ahead of the caboose that became her home on wheels.

"Udderly" Clean Milk

As a bacteriologist for the Arizona State Laboratories, civil engineer **JANE H. RIDER** (1889–1981) found that the cause of most infant deaths in Arizona in the early 1900s was unsanitary milk. In rural areas, milk cows would stand hip-deep in 'dobe holes—the holes left in the ground after digging out material to build adobes—to drink water.

The problem was that the cows' owners didn't wash the udders before milking. Dirt and germs from the stagnant water were getting into the bucket along with the milk. Jane publicized the link between infant mortality and contaminated milk, and finally convinced the dairy industry to pasteurize milk.

Jane was also an advocate for clean drinking water. Harmful waste and sewage disposal, particularly from mining, polluted so many streams that pure water was hard to find. As a life-long supporter of Arizona water works, Jane continually fought industrial pollution.

The tireless Jane was remembered for her sense of humor, which, she claimed, opened many doors to a pioneering woman engineer.

Jane Rider fought the popular practice of selling food items in open barrels at grocery stores where goods were easily infested with bugs and rodents, and helped establish sanitary food displays. She also put a stop to the practice of selling substandard food, medicine, and cosmetic products in Arizona before the Pure Food and Drug Act existed. Some of the hair dyes sold contained such caustic ingredients they caused convulsions!

When Jane Rider was 15 she decided to follow in the footsteps of her father who was also an engineer. She was the first woman to graduate with a degree in engineering from the University of Arizona. For the next 50 years, Jane improved the health of her fellow Arizonians by leading the cause for better sanitation, cleaner water, and more effective hospital design.

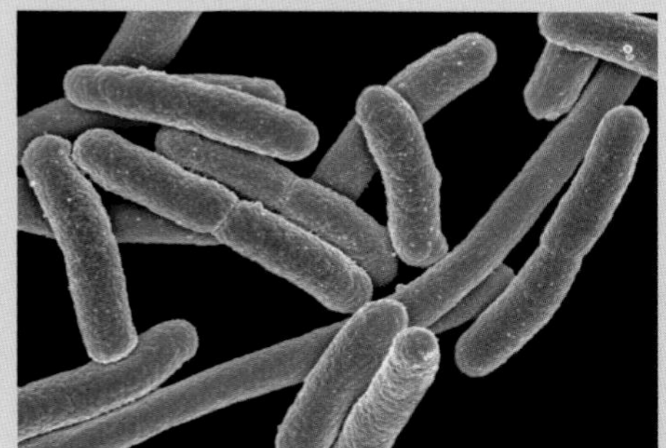

Escherichia coli Bacteria, commonly known as *E. coli*, can cause food poisoning when found in above-average numbers.

Avoiding Tummy Trouble

Bacteria are everywhere, even in our own intestines. Most bacteria are not harmful. But there are at least 250 known foodborne diseases caused by bacteria, viruses, or parasites. Bacteria, including *Clostridium botulinum*, strains of *Escherichia coli*—or *E. coli*—and *Salmonella*, can make you sick.

As prevention:

- Wash hands with hot, soapy before preparing food and after handling raw meat, poultry, or eggs.
- Wash utensils and kitchen surfaces with hot soapy water after food preparation and after handling raw meat, poultry, or eggs.
- Cook meat (especially hamburgers!), poultry, and eggs thoroughly.
- Refrigerate leftovers within two hours of cooking.
- Wash fruits and vegetables thoroughly.
- Drink pasteurized milk and apple cider.

Farming the Natural Way

Good farming is a constant balance between what's best for farmers to grow commercially successful crops and what's best for the environment.

Bugs, fungi, and parasites can cost farmers millions of dollars in crop damage. Herbicides, pesticides, insecticides, and fungicides are often used to control the spread of weeds, insects, and disease.

But if not properly applied, these "'cides" can harm the environment. Some farmers are turning toward more natural ways to control pests on crops, lawns, and gardens, and in commercial greenhouses.

They're also using natural biological processes to treat wastewater, and to grow crops that use just the right amount of fertilizer to yield a robust harvest. Guess what? Engineers help farmers figure out how to do this!

Worms to the Rescue

Fish eat worms. But will worms eat fish poop? It's an important question for the aquaculture (farming of plants and animals that live in water) industry. One of the greatest costs for Blue Ridge Fisheries—the largest producer of tilapia in the U.S.—is disposing of the wastewater from the fish tanks, which contains fish poop and uneaten fish food.

Fish waste is often used in fertilizer but must be treated with bacteria to break it down to the desired material. That's expensive. To get some creative suggestions, Blue Ridge Fisheries turned to **LORI S. MARSH** (b.1956), an expert on aquaculture and vermiculture—the production of worms.

Would worms work? Lori took a few worms and fed them fish waste straight, with nothing else mixed in. "They loved it! So, I built some pilot worm beds and broadened my experiment. That worked too!" says Lori.

Worms thrive on fish waste. They can be sold for bait or fish food. And worm waste can be used as fertilizer without further breakdown. "This solution is so perfect and cyclical. It shows engineering is not just about machines, it's also about living and biological systems."

Lori Marsh is holding vermicompost—compost made by worms—in front of the trammel screen used to separate worms from compost. The worms are then put back into the worm bed or sold. The compost is sold to greenhouses to use in potting mixtures.

The Good, The Bad, and The Wiggly

Microscopic worms, called nematodes, are used extensively as biopesticides—which are living organisms such as bacteria, fungi, viruses, and parasites that control pests. Bugs eat the nematodes and then die from the bacteria that the nematodes carry. Different species of nematodes destroy different species of bugs, so not just any nematode will do.

The question is how do you make sure the nematodes don't die as they're sprayed onto crops? Agricultural and biological engineer **JANE PATTERSON FIFE** (b.1975) tackled the problem. She wondered whether conventional spray equipment used for chemical compounds could also be used safely for bio-organisms that must be delivered "live" to do the job. If so, it would save farmers from having to buy new spray equipment.

Jane tested spray nozzles, and pressure and temperature settings, and used computerized modeling to discover which system nematodes preferred. She found that cone-shaped nozzles and low-temperature pumps, commonly used for chemicals, would also work for nematodes. Problem solved!

Entomopathogenic nematodes are used as a biological pesticide against a wide variety of insect pests. For example, they can be used to destroy grubs that cause millions of dollars a year in damage to lawns and golf courses.

Jane Fife is working here with a Phase Doppler Particle Analyzer, an instrument that measures spray droplet size and velocity based on light refraction by droplets as they pass through laser beams. The size of pesticide droplets passing through spray nozzles is important because very small droplets are more prone to drift or can be inhaled by humans.

Conventional pesticides are made from synthetic materials that kill pests. Biopesticides and biochemical pesticides are made from natural, non-toxic substances that control pests. For example, enticing fragrances made from plant extracts can attract insects into traps.

If your houseplant needs help in a hurry, you might try putting a little canola oil or baking soda on the insect-infested area.

Just Enough is Just Right

"Many of us go into agricultural engineering because we want to protect the environment," says agricultural engineer **MARY LEIGH WOLFE** (b.1957). Mary Leigh teamed up with soil scientists and government agencies to develop a practical index tool to help Virginia's farmers determine how much phosphorous can safely be added to the soil.

Farmers input data—such as soil characteristics, the slope of the land, the distance from land to water, and types of fertilizers—into Mary Leigh's computer spreadsheet. The spreadsheet then calculates the risk of high phosphorous levels and tells the farmer how much fertilizer she can apply.

While the final index is still pending state regulation, Mary Leigh is confident that Virginia will soon have a new tool to effectively manage phosphorous. Improved water quality statewide will be the result.

Mary Wolfe obtains soil samples using a soil probe. The samples are then analyzed for phosphorus content. This is an important step in determining how much liquid manure can be applied to cropland using a big gun (on right) without causing water quality problems.

Healthy Drinking Water

When civil engineer **WENDY DIMBERO GRAHAM** (b.1960) studied the impact of citrus crops on Florida's groundwater—the water supply for many people—she discovered there were unhealthy levels of nitrates getting into drinking water. The chemicals farmers use in fertilizer and livestock feed can leach into the earth, filter down through soil and rock layers, and mix with the groundwater below. This makes groundwater unhealthy to drink.

Wendy worked with farmers, scientists, and state regulators as she researched and tested soil and plant specimens over a long period of time. Her constant monitoring and analysis helped farmers reduce the amount of nitrates they were using. Eventually the drinking water returned to its original high quality.

"Many people are often surprised by how willing farmers are to protect the environment," says Wendy. "They realize how sensitive the environment is and how important a healthy environment is to agriculture."

Wendy Graham also participated in studies involving phosphorus in water runoff from beef cattle ranches, and fertilizers leaching into springs feeding the Suwannee River in Florida.

Livestock Needs Love Too!

Agricultural and biological systems engineer **EILEEN FABIAN WHEELER** (b.1957) agrees that engineering helps people lead better lives. She has a caveat, though. "Engineering helps animals lead better lives as well."

As a girl, Eileen loved horses. She still does. She chose engineering as a career to combine her love of animals with her fascination about how things work.

As a horse lover, Eileen knows they can develop respiratory problems. As an engineer, she studied horse stables and enclosed riding arenas to see if she could improve the air quality.

Eileen measured temperatures and humidity, dust, and the gas in the air, but discovered the major problem in modern horse stables was simply a lack of openings for fresh air.

The dust was caused largely by bales of hay stored in barns where the horses lived. To help the horses breathe more easily, Eileen recommended providing a fresh air opening to each stall and storing hay, and its dust, away from horse living quarters.

Eileen finds the breadth of her work fascinating and enjoys the collaboration with colleagues from other fields. "I'm always learning something new." She thinks engineering enables women to become more independent intellectually, financially, and socially.

Eileen Wheeler has made livestock more comfortable and less stressed in many ways. For example, she helped horses live in stables with fresher air, and provided comfortable barn environments for chickens and calves. She's also promoted wind tunnel ventilation in livestock barns to help animals keep cool in hot temperatures.

Powered by Corn

Commercially produced ethanol is made from corn kernels processed with microorganisms. But using corn kernels for ethanol decreases the food supply for people and animals that would otherwise be eating it.

Biosystems and agricultural engineer **SUE E. NOKES** (b.1960) and two colleagues wanted to see if ethanol could be made from the corn plant's inedible leaves and stems. Sue's team mixed corn biomass with *Clostridium thermocellum*, the bacteria found in cows' digestive systems. Ethanol was produced, but only for a short time until the ethanol killed the bacteria. Sue's team then discovered a way to separate the ethanol from the fermentation process so the ethanol would not kill the cells.

Though less ethanol was produced than with the process using corn kernels, Sue's group was the first to prove that ethanol could be made using corn leaves and stems.

Sue Nokes (above right) is researching ethanol with the help of her students, including Mari Chinn (above left). Sue has won awards for her creative style of teaching. "When we talk about 3-D enzymes in class, we build them out of play dough. I love encouraging students to break out of the mold when they're puzzling out problems. Creative ideas add so much to collaborative research."

Tools to Grow By

From the sunny picture of farm life often depicted in commercials, movies, and TV, most people wouldn't guess that farm-related injuries affect an estimated 33,000 children each year in the U.S. Agriculture is one of the most hazardous industries.

Much has been done to make agricultural equipment safer to use. Farm injuries and deaths have been greatly reduced, especially in the last 20 years when injuries have decreased by almost 50 percent.

"I think that I shall never see
a billboard lovely as a tree.
Perhaps, unless the billboards fall,
I'll never see a tree at all."

Ogden Nash
Humorist and Poet (1902–1971)

Awatif Hassan's semi-automatic machine carried small trees, one at a time, between two plates that penetrated the ground as a tree was spot planted. Two operators—a tractor driver and a tree planter—were needed to maneuver the machine through the forest.

Tree Planter Extraordinaire

Most people don't think of trees as a crop—but they are. Millions of trees are harvested every year by the construction and paper industries.

In the 1970s, agriculture engineer **AWATIF E. HASSAN** (b.1937) worked on a machine that automatically and uniformly planted trees. Her semi-automatic spot planter could move easily around stumps in the forest. "At the time, labor was in short supply. Hand-planting was often unsuccessful because tree seedlings can die if their roots are exposed to air for more than 10 seconds," Awatif explains.

In 1983, Awatif unveiled an advanced semi-automatic spot planter. Awatif recalls, "the planters were never commercially successful because the economy changed and labor was cheaper than buying the machine, but it was a great project, and it was fun while it lasted."

Roll Bars for All

The number-one cause of farm fatalities is overturned tractors. While working at the University of Iowa, agricultural engineer and safety expert **CAROL J. LEHTOLA** (b.1951) launched the TRAC-SAFE (Tractor Risk Abatement and Control) project to install rollover protective structures.

"Many farmers keep tractors for as long as 50 years, and updating them with new equipment was a challenge to say the least," commented Carol. The program was so successful that TRAC-SAFE was adopted nationally in the mid-1990s.

Carol Lehtola (right) is also childproofing farms and helping family farmers follow safety measures when their kids are doing chores. She was a primary advisor for a project to develop agricultural safety guidelines for kids. Her guidelines match tasks appropriate to a child's physical and mental development so that work can be an enjoyable learning experience with minimal danger.

Farmers Share the Road

"I grew up on a small farm and was always curious about how machines worked," says agricultural engineer **MELANIE WILLIAMS HARKCOM** (b.1957). That's what drew me to agricultural engineering and developing farm equipment." Melanie now designs, develops, and tests farm machinery and equipment.

When building new equipment, Melanie Harkcom (left) emphasizes, "it's not just about drawing a design. There are many more details involved than most people think."

Merging farm practices with the hustle and bustle of encroaching suburbia has created special challenges. As roads and highways become more congested, restrictions have been placed on the size and width of farm equipment to ensure safer driving conditions for all.

Melanie recently worked on the lateral transporter, a special trailer that pivots wide equipment, like hay mowers, to a vertical angle so the equipment is in a position that's slim and safe enough to be transported on roads.

Modern Households

Women have always been a driving force in their homes—making the world a better place for their families and children.

Today, there are more women in the workforce. There are more women engineers, too, who continue to improve our quality of life but in much broader ways. Women engineers can now reach into hundreds, thousands, even millions of modern households through their life's work.

Women engineers design safe and comfortable homes, business spaces, and community centers. They invent or improve timesaving devices so that people can have more time with their family and friends.

Women engineers make sure our drinking water is clean and safe. They also see that sewage from cities and industries is treated to improve public health and make sure the natural habitats around us stay clean.

Water pollution and sewage control plant, Canada, aerial view.

A Good Roof Over Our Heads

Structural engineers create the backbones for all sorts of buildings: our homes and offices, places of worship, and schools. They design foundations, beams and columns, and roofs for everything from small houses to huge high-rises.

Engineers also create buildings to be artistic expressions of the occupants inside. Sometimes the buildings are designed to blend with the surrounding environment. In other cases, they stand as icons, like the Empire State Building or San Francisco's Transamerica pyramid.

Engineers design buildings strong enough to protect occupants from nature's forces—such as hurricane winds, earthquakes, or heavy snowstorms—to make sure we have a safe and secure roof over our heads.

Saw-Teen See is managing the structural design of the 90-story Shanghai World Financial Center (right) that houses office, retail, and hotel space.

REACH FOR THE SKY

When complete, Shanghai's World Financial Center in China will stand over 1,600 feet high with 90 stories of office, retail, restaurant, and hotel space topped by an observation deck and museum. Structural engineer **SAW-TEEN SEE** (b.1954) is managing the building's design.

"Not only will the skyscraper be one of the world's tallest buildings," says Saw-Teen, "but its internal steel structure has been designed so the building can withstand huge seismic forces from an earthquake and enormous wind forces from an intense typhoon!"

The most amazing thing about the center is that a shorter, smaller building had been planned for the same space. "In fact, the foundation had already been built when the owners decided to make the skyscraper bigger," says Saw-Teen.

She and her team of engineers worked together to design a lighter structural system so a larger, taller building could be constructed without changing the existing foundation.

Saw-Teen takes her work seriously because errors could cost lives. At the same time, she and other structural engineers, working with architects and building owners, look beyond pure calculations to create a pleasing, well-functioning space. "It's a building we're designing, not just a structural problem to be solved," she says.

> *"Engineering is very rewarding because you can see and touch what you've created."*
>
> —SAW-TEEN SEE

ABOVE: SAW-TEEN SEE DESIGNED THE ROCK AND ROLL HALL OF FAME IN CLEVELAND, OHIO.

SAW-TEEN (LEFT) SIGNS HER NAME TO A SUPPORTING BEAM DURING CONSTRUCTION OF THE BALTIMORE CONVENTION CENTER.

CALIFORNIA CASTLE

In 1919, newspaper tycoon William Randolph Hearst hired architect and civil engineer **JULIA MORGAN** (1872–1957) to design a home on his central California property. Julia had worked on family estates and public buildings—including the YWCA conference center at Asilomar—for Phoebe Hearst, William's mother. But the millionaire wanted something more extravagant for himself.

The result, Hearst Castle, is a 127-acre complex with a 165-room house, guesthouses, swimming pools, gardens, tennis courts, and, at one time, even a zoo! For over 20 years, Julia supervised the building of Hearst Castle.

Hearst treasured Julia's skill and adaptability. When he asked for changes—some as extreme as adding another level to a completed building—Julia re-engineered the building to meet his wishes. She designed rooms to accommodate medieval and Renaissance church ceilings and other artifacts Hearst bought in Europe and had shipped to California.

DURING CONSTRUCTION OF HEARST CASTLE (ABOVE), WILLIAM RANDOLPH HEARST AND JULIA MORGAN (ABOVE, LEFT) COLLABORATED ON IMPROVEMENTS TO THE DESIGN. WHEN COMPLETE, BOTH COULD SAY HEARST CASTLE MET THEIR HIGH STANDARDS.

Architect? Engineer? What's the Difference?

Many think **JULIA MORGAN** was an architect, someone who designs the exterior and interior of a building. But Julia was also a civil engineer trained to design a building's support system and foundation.

After earning a civil engineering degree at the University of California, Berkeley, Julia studied architecture at the most highly regarded architecture school of her time, the École des Beaux Arts in Paris. At first, she was refused admission because she was a woman. Julia persisted and graduated in 1902. By 1904, she owned and managed her own firm, designing homes and public buildings and overseeing their construction.

Before retiring in 1951, Julia designed and built over 700 structures including churches, schools, hospitals, women's clubs, homes, and university buildings. Sixteen are listed on the National Register of Historic Places, and two are National Historic Landmarks: the conference center at Asilomar and Hearst Castle.

DOLLARS AND CENTS

Contractors need to know the price of supplies, labor, and materials to estimate the cost to build a skyscraper, home, bridge, or any other project. For 30 years, civil engineer **ELSIE EAVES** (1898–1983) put construction statistics—including wages, material costs, and construction activity reports—into the hands of the industry. She did this at *Engineering News-Record,* the top magazine covering construction. To be sure the information stayed up to date, Elsie managed a team of nearly 150 field and pricing reporters and 25 office staff.

Her career spanned building boom times and depression years. In fact, during the Great Depression, her inventory of planned and needed construction projects in the United States was used to support federal funding that sparked new construction. In the 1940s, Elsie created the nation's official report of construction work that could proceed at the end of World War II.

ELSIE EAVES IS PICTURED HERE IN 1920, THE YEAR SHE EARNED HER CIVIL ENGINEERING DEGREE. SHE ENJOYED A NEARLY 50-YEAR CAREER INCLUDING 30 YEARS WITH THE PUBLICATION *ENGINEERING NEWS-RECORD.*

RUTH GORDON'S EARTHQUAKE SAFETY WORK WAS A FAR CRY FROM ONE OF HER FIRST JOB OFFERS. IN THE 1940S, A BATHING SUIT MANUFACTURER WANTED TO SIGN HER ON SO HE COULD SAY A WOMAN ENGINEER DESIGNED THE COMPANY'S SWIMSUITS. RUTH TURNED DOWN THE OFFER BECAUSE THE MANUFACTURER HAD NO INTENTION OF PAYING HER AN ENGINEER'S SALARY.

WATCHING OUT FOR THE PUBLIC

What goes up can surely come down due to natural disasters. Structural engineer **RUTH V. GORDON** (b.1926) spent over 30 years making sure our most important buildings could withstand these natural forces.

After several years in structural design, Ruth became a senior structural engineer for the State of California. "I reviewed plans for schools and hospitals to determine if the buildings would protect occupants from wind, fire, and earthquakes," says Ruth. "I also made sure hospitals could be operational after a natural disaster."

Ruth traveled a 200-mile radius to check construction sites and explain earthquake-resistant building techniques to contractors. "I really loved structural engineering," she says, "especially out in the field."

She founded her own engineering firm in 1984. In 1989, after a 7.1 magnitude earthquake rocked Northern California, Ruth made sure the extensive San Francisco General Hospital complex and other key buildings were safe for use. Both before and after that big quake, Ruth gave many talks on earthquake safety to the general public.

Ruth is the mother of three, all working in scientific fields, and a grandmother of four.

HOME IMPROVEMENTS

For centuries, women have kept the home fires burning: cooking meals, sewing clothes, raising children, and cleaning house. So who would be better than a woman to engineer more comfortable, safe, and convenient homes?

Consider this: Before air conditioning, people in hot climates sweltered in the summer. Before washers, driers, and electric irons, doing the family's laundry was a several-day ordeal. Before electric and gas stoves, cooking over a wood-burning stove was tricky. Before vacuum cleaners, rugs had to be hung outside and beaten with a broom to get rid of the dirt.

It's easy to forget just how valuable engineers' contributions have been in making our homes nicer places to live—and so much less time-consuming to maintain.

"Engineering is the art or science of making practical."

SAMUEL C. FLORMAN
ENGINEER AND AUTHOR (B.1925)

LEFT: THIS "KITCHEN OF THE FUTURE" WAS DESIGNED BY KELVINATOR IN 1957. BACK THEN, DESIGNERS IMAGINED A TYPICAL KITCHEN IN THE 1960S AND 1970S WOULD INCLUDE A REFRIGERATOR POWERED BY ATOMIC ENERGY RAYS, AN ULTRASONIC DISHWASHER, AND A PUSH BUTTON CONTROL TO MAKE APPLIANCES DISAPPEAR UNDER COUNTERTOPS.

CENTER: A STATE-OF-THE-ART TOASTER, IN 1930, THAT IS.

RIGHT: A VACUUM CLEANER OF THE MID-20TH CENTURY.

CHEAPER BY THE DOZEN

With a family of twelve children, **LILLIAN MOLLER GILBRETH** (1878–1972) could have been quite busy without a career in engineering. The Gilbreth family was the subject of the book and movies, *Cheaper by the Dozen*.

Lillian and husband Frank ran Gilbreth & Company, a firm that studied manufacturing tasks to find ways to save money and reduce workers' injuries. When her husband died in 1924, Lillian ran the company even though her youngest child was still a toddler.

One of Lillian's most famous studies benefits anyone who uses a kitchen today. She measured 4,000 women and observed them using a typical kitchen. She noted how far they bent down to open oven doors, how far they reached up for stored items, and the total distance they traveled from appliance to appliance to make one coffeecake.

Lillian took the information and in 1929 designed a "Kitchen Practical" to demonstrate her more efficient, comfortable arrangement of kitchen work surfaces, stoves, and sinks.

In the 1950s, Lillian expanded her studies to find ways for disabled people to cook their own meals. Her principles for an efficient kitchen are used today in basic kitchen design. Another reminder of Lillian's work can be found in most kitchens: the trashcan with a foot pedal to raise the lid, hands-free!

ABOVE: IN LILLIAN GILBRETH'S "KITCHEN PRACTICAL" OF 1929, APPLIANCES AND WORK SURFACES WERE LOCATED SPECIFICALLY FOR MORE EFFICIENT FOOD PREPARATION.

LEFT: LILLIAN HELPED REVOLUTIONIZE INDUSTRIAL ENGINEERING PRACTICES AND INFLUENCED THE LAYOUT OF TODAY'S KITCHENS, ALL WHILE RUNNING A COMPANY AND RAISING 12 CHILDREN!

Ironing Out the Details

In the 1920s, when electrical engineer **LOTTYE E. MINER** (1904–1989) was in college, electric appliances were just making their appearance in home kitchens. She believed the use of electric appliances would expand, and wanted to make her mark as both a woman and an engineer in the appliance industry.

As an appliance development engineer with both Westinghouse and Coleman Lamp and Stove, she experimented with the first thermostats in irons, waffle irons, and toasters to find the best wattage—or power level—for the appliances to heat up without burning. She once took several test irons to a friend's house and ironed cotton cloth to see which worked best.

Later in her career, Lottye Miner and her husband formed an electrical engineering company to expand electricity to rural America and bring electrical power to undeveloped areas of Lebanon, Brazil, and Pakistan. Lottye continued to run the firm after her husband's death.

Rays of Sunshine

MARIA DE TELKES (1900–1996) believed that the sun's unlimited power could be used for energy instead of coal or other fossil fuels. She hoped that some day, solar energy would power homes, businesses, and appliances.

During World War II, Maria developed a solar-powered device installed on life-rafts to convert seawater into drinking water. In 1949, Maria and architect Eleanor Raymond built the first solar-heated house, located in Dover, Massachusetts. She continued to participate in solar house construction projects through the 1970s.

Maria is probably most famous for inventing a solar stove that could reach 350°F, hot enough for baking. She thought the solar stove would benefit countries with abundant sunshine and little fuel or electricity. Today, scientists continue to introduce solar stoves similar to Maria's design to people living in Kenya and other developing countries.

Maria de Telkes (left) proudly shows off the "Meritorious Contribution to Engineering" award she received from the Society of Women Engineers for her design of a distilling system to use solar heat to convert seawater to drinking water. With Maria (left to right) are Katharine Stinson (see page 145), Lillian Murad, and Beatrice Hicks (see page 162).

Better Than Ever

Engineers continue to improve household products. A top focus today is developing energy-efficient lighting and appliances to reduce power needs.

For example, compact fluorescent lamps (CFLs) now come in styles to fit most light fixtures. CFLs use one-third the electricity of regular light-bulbs and last up to 10 times longer. In fact, if every U.S. home replaced five frequently-used light-bulbs with CFLs, the output of 21 power plants would be saved!

Refrigerators built with power-saving compressors, better insulation, and precise temperature controls use almost half the electricity of older models. Energy-efficient televisions shave 25 percent off the electricity demand of older styles. And today's computers use a whopping 70 percent less energy than their predecessors thanks to power management features such as "sleep" modes.

It Is the Humidity!

As a child, mechanical engineer **MARGARET INGELS** (1892–1971) wondered why moisture collected—or condensed—on the inside of a cold window. Condensation would become an important part of Margaret's life. Her chosen field, air conditioning design, involves the control of indoor air's moisture content (also known as humidity), temperature, and rate of motion.

Early in her career, Margaret researched air quality and its physical and psychological effects on people. In one project, she considered children's health and daily attendance when schools used different types of ventilating systems. Margaret also developed a temperature scale that related human comfort to not only air temperature but humidity and air motion, too. She proved scientifically what might seem obvious today: that people feel more comfortable in a dry or breezy climate than in a humid climate, even if the thermometer reads the same temperature in both.

Later, Margaret joined Carrier Engineering Corporation, a pioneering air conditioning firm. She designed air conditioning systems for the manufacturing of textiles, tobacco products, and other materials sensitive to temperature and humidity.

Margaret is most well known for designing and promoting air conditioning for home use. Many people today take the air conditioning in their homes for granted. But in Margaret's time, air conditioning a residence was a new idea. She educated the public through lectures and magazine articles that made air conditioning easy to understand. Over the course of her crusade, Margaret spoke to an estimated 12,000 people, wrote over 50 articles on air conditioning, and was interviewed on television.

Margaret Ingels devoted her career to air conditioning design. She's shown here in 1917 at the University of Kentucky while pursuing her mechanical engineering degree, holding a piece of machinery she invented.

Margaret Ingels was a great friend of Willis Carrier, known as the "father of air conditioning." She wrote his biography.

Her efforts led to the widespread understanding, acceptance, and use of air conditioning in the United States. Margaret also wrote *Petticoats and Slide Rules* in 1952, a tribute to early women engineers.

Today, heating, ventilating and air conditioning (HVAC) engineers use the latest technology to make living and working in extreme climates—very hot or very cold—much more tolerable.

MARTHA THOMAS HELD OVER 20 PATENTS FOR IMPROVING LIGHTING TECHNOLOGY AND MANUFACTURING. WE NOW KNOW THAT IF EVERY HOME IN THE U.S. REPLACED JUST ONE INCANDESCENT BULB WITH A COMPACT FLUORESCENT LAMP, POWER PLANT OUTPUT WOULD BE REDUCED SO MUCH THAT IT WOULD BE THE SAME, IN TERMS OF AIR POLLUTION, AS TAKING ONE MILLION CARS OFF THE ROAD!

BRIGHTER IS BETTER

Most light-bulbs are incandescent, a huge improvement over kerosene lamps and candles of old. The problem is that incandescent bulbs waste energy. Only 10 percent of the energy drawn by a bulb produces light we can see. The rest is given off in heat. When you touch a lit light-bulb: ouch!

Compact fluorescent lamps, or CFLs, require less electricity because they emit mostly visible light. That helps save our energy resources. CFLs last up to 10 times longer than regular light-bulbs, too. CFLs are way cool, literally! They stay cool to the touch.

Early fluorescent bulbs gave off an eerie, green light. Imagine trying to read by that! **MARTHA J. B. THOMAS** (b.1926) developed a white phosphorus powder to coat the inside of fluorescent tubes. With this coating, the lamps give off a light more like natural daylight. That made Martha a real eye saver!

ANATOMY OF A LIGHT-BULB

The design of the light-bulb has changed little since 1879. But the materials used to make light-bulbs have improved.

Tungsten is commonly used for filaments, the tiny coiled wire that glows inside a light-bulb. Tungsten won't melt until it reaches over 6,300°F. Most other metals would turn to goop inside the hot bulb.

Metallurgist **EDITH PAULA CHARTKOFF MEYER** loved to study metals so much that she said she "must have had a metal lining to her skin." In the 1920s, Edith researched new ways to use tungsten by blending it—or alloying it—with other metals.

Molybdenum, called "moly" in the materials world, is another versatile and sturdy metal. It's mixed with steel to make products that stand up to heavy use and high temperatures—such as tools and parts for furnaces and cars. Molybdenum is also used for filament supports in light-bulbs.

One of the world's top experts in molybdenum was metallurgical engineer **JANET Z. BRIGGS** (1912–1974). Her research led to an expanded use of molybdenum in manufacturing metal parts.

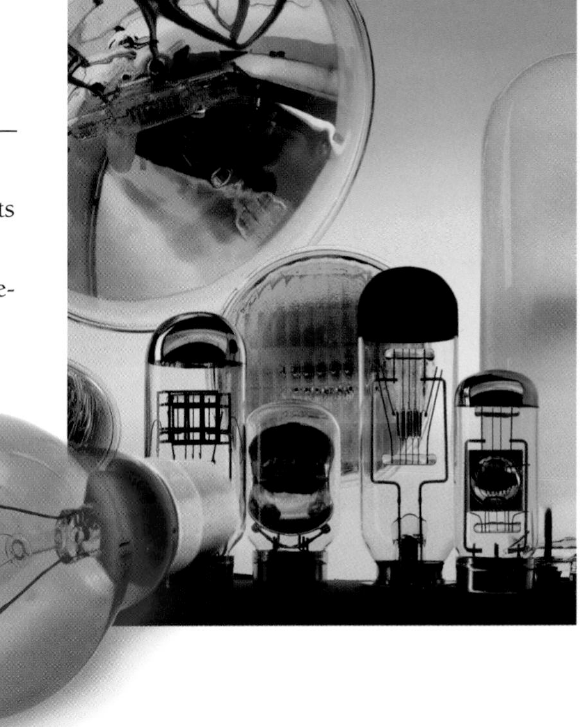

Real Life Chem Lab

Chemical engineer **LINDA LAFFLER BOLTON** (b.1967) has designed chemical processes and equipment for a flooring manufacturer and a company supplying chemicals to other manufacturers.

Her main challenge is balancing product needs with available budget. "I've never been given an open checkbook and been able do what I wanted *carte blanche,*" she says.

In one project, Linda was asked to produce a polymer additive in liquid form. "This additive had been sold as a solid," explains Linda. "The change was proposed to make the product easier to use and to increase sales." The only catch: the additive degraded at high temperatures, yet had to be heated to be both purified and kept as a liquid.

She found a way to achieve this tricky goal, but the required equipment was quite large and expensive. "So, I designed the equipment to be used for other products and processes so the cost could be shared," she says. Her work earned her the prestigious Henry D. Wright Achievement Award for Outstanding Process Development and Design.

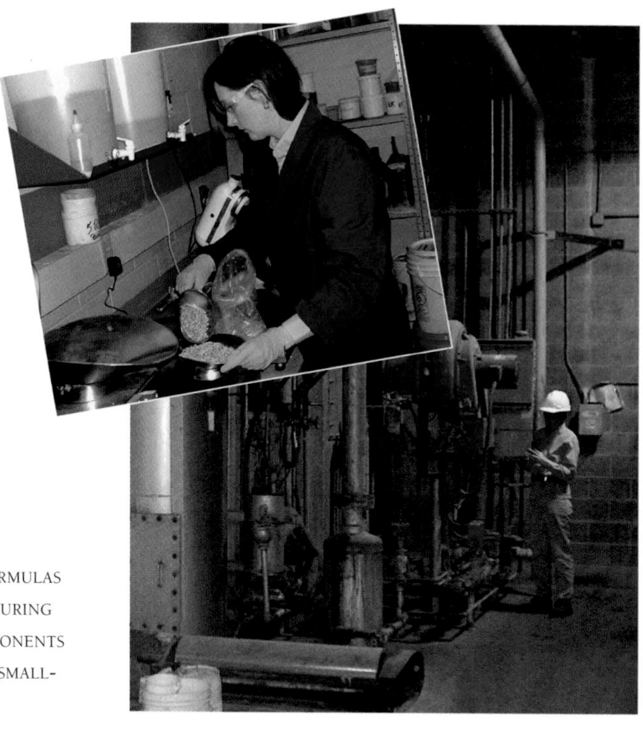

Linda Bolton must evaluate new product formulas before they are introduced to the manufacturing process. She does this by measuring the components (ingredients) and developing the process in small-scale experiments.

Sweet Cents

"There are an extraordinary number of ingredients in Downy® fabric softener fragrances. Combine this with all the facets of manufacturing, packaging, and shipping the product. I was really challenged," says civil engineer **NANCY E. URIDIL** (b.1951).

Nancy has worked on a wide variety of commercial and household products from toilet paper and laundry products at Procter & Gamble to 23 cosmetic brands at Estée Lauder, and now faucets at Moen.

"I started out building water systems and subdivisions, then building products, brand equity, and technical capabilities," says Nancy, who has done it all.

Helping make Downy number one in the market place was one of her favorite jobs. "Consumers loved our product and package improvements. Before long, we were selling Downy so fast that our CEO flew in to see what we were doing right. That visit marked a turning point in my career."

Nancy Uridil enjoys making top brands, from luxury cosmetics at Estée Lauder to beautiful faucets at Moen.

Tap Technology

Water is our most basic need. The average American uses almost 100 gallons of water each day for drinking, food preparation, cleaning, and irrigation. Clean, abundant water is a necessity for farming and industry as well, but there's a lot of engineering "upstream" of the tap.

Many different types of engineers work together to design treatment plants and pipelines that provide clean drinking water. As a result, waterborne diseases caused by bacteria and viruses have been nearly eliminated in developed nations.

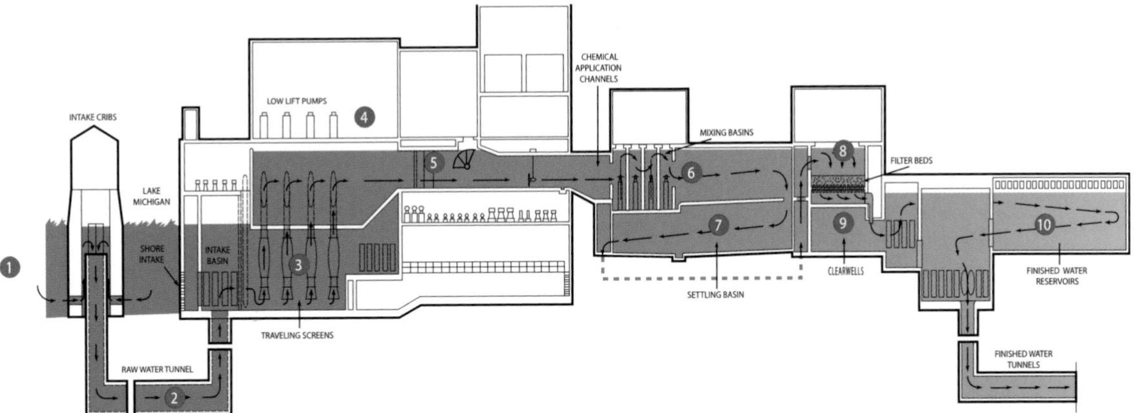

This drawing shows each step of Chicago's drinking water treatment process from Point 1, where source water is drawn from Lake Michigan, to Point 10, where clean, safe water flows to household and business taps as well as fire hydrants.

Barbara Fox met with the public, especially students, whenever possible to answer their questions about the quality of their drinking water and the way it was delivered. At left, she shows members of the public how the water system works during "Water Appreciation Week in Chicago," in 1976.

Urban Waterworks

Over five million urban and suburban consumers rely on the City of Chicago's water system. Two intake cribs, two of the world's largest water purification plants, 12 pumping stations, and 65 miles of tunnels supply their high-quality drinking water.

Civil engineer **BARBARA FOX** (b.1935), the first woman hired as an engineer by the city's Department of Water, rose to the position of water engineer in her 36 years of working there.

"I enjoyed the interactive nature of my work, conferring with other engineers, scientists, and government officials in Chicago and around the country," says Barbara. "We all shared the goal of providing the best-quality water for consumers no matter what challenges we encountered."

One of the greatest challenges came in the early 1990s when a tiny critter, the non-native Zebra Mussel, threatened Great Lakes water supplies. "The mussels upset the ecosystem and blocked water intakes," says Barbara. She and other engineers and scientists pondered expensive, complex solutions, but didn't overreact. Good thing! Drastic measures weren't needed. Water department workers simply inspect and clean intake screens and treatment plant basins, and the situation remains stable today.

Tasty Water

Water industry engineers must plan ahead so their water can continue to meet drinking water standards—even if the regulations become more stringent. While the drinking water provided by the Contra Costa Water District, just east of San Francisco, is fine by today's standards, civil engineer **SAMANTHA SALVIA** (b.1974) is looking ahead to possible changes in drinking water requirements.

"The challenge is, our utility, which serves about half a million people, gets its water from the Sacramento-San Joaquin River Delta," explains Samantha. "The Delta has water quality problems. For example, during droughts, salty water from San Francisco Bay can reach our intakes."

Samantha and her team are moving the water district's intake to a location further upstream with better water quality. That way, when the river water is treated, it will yield drinking water that meets more stringent regulations expected in the future.

"Changing the location of a drinking water intake involves a lot more than designing pipes and pumps," she says. "We must understand how the project will interact with the ecosystem, other cities, and agriculture."

Samantha Salvia works for the Contra Costa Water District in California, which supplies drinking water to over 500,000 customers. Outside of work, Samantha Salvia is a member of Fury, the San Francisco Bay Area's Women's Ultimate Frisbee team. She has competed in national and international championships.

Worldwide Reach

Civil engineer **KRISTY SCHLOSS** (b.1963) has been President and CEO of Schloss Engineered Equipment, Inc. for 16 years. It's a third-generation firm that designs and manufactures water and wastewater treatment equipment used in 36 countries around the world.

Kristy combines her engineering knowledge with business and financial savvy as she travels internationally to meet with customers, other engineers, and high-level diplomats.

"I am proud that every device installed in a town, village, or even a large city provides clean drinking water and a healthier environment," says Kristy. "As a result, communities experience better health and sustainable development."

Kristy Schloss's experience in the international business world led to her appointment to the Federal Reserve Bank Board, Denver Branch, and the Department of Commerce's Environmental Technologies Trade Advisory Committee. Kristy serves on a number of other boards concerned with international trade and education.

Far right: When clean water is readily available, girls in developing countries are free to attend school because they are not needed to tote water several times a day from nearby streams.

The Great Outdoors: A Day in the Life of Debra Kaye

The jagged, snow-topped Sierra Nevada Mountains, sparkling Lake Tahoe, and color-dotted alpine meadows are more than the beautiful sights of Washoe County, Nevada. They serve as workplace and playground for civil engineer Debra Lynn Kaye (b.1966).

Debra is the operations and maintenance manager for the Truckee Meadows Water Authority. She's responsible for delivering clean, safe drinking water to 360,000 customers in Reno, Sparks, and the surrounding area.

Although she's a manager, chances are you won't find Debra sitting behind a desk. Each day's new, interesting challenges take her to many points across the picturesque region.

"I wear a lot of hats because this is a small organization," says Debra. "That's one of the things I enjoy about working in the water industry."

DEBRA MAY START HER MORNING MEETING WITH THE WATER COMPANY'S ELECTED BOARD MEMBERS. "THEN, WITH ONE PHONE CALL, I HAVE TO CHANGE INTO JEANS AND WORK BOOTS TO CHECK OUT A CONSTRUCTION PROJECT. RIGHT NOW, WE'RE BUILDING A SPILLWAY FOR THE HIGHLAND CANAL, ONE OF THE WATER SOURCE SUPPLIES TO THE CHALK BLUFF TREATMENT PLANT."

THE SIERRA NEVADA MOUNTAINS NEAR RENO, NEVADA.

"My daughter drinks this water. My mother drinks this water. I know my community is protected by what I do. That's the most satisfying part."

—DEBRA LYNN KAYE

Before becoming an engineer, Debra was a hands-on water treatment plant operator. "I knew I could combine my practical experience with an engineering degree to better design and operate water facilities," she says. She works hand-in-hand with plant staff to develop operational strategies.

"Our water has a river source, so a storm can change water quality. We're always troubleshooting, especially when it rains or snows," says Debra.

"By working as an engineer, I believe I'm a positive role model for my daughter," says Debra. She is shown here with her mother, Robin Collins, and her daughter, Alexis.

There is a sport for every season in the Truckee Meadows area where Debra lives. "In the winter, I love to snowboard. When the weather turns warmer, I take every opportunity to slalom water ski, relax on a houseboat, ride my Harley-Davidson motorcycle, or combine outdoor activities. Spring is my favorite time of year because, when the weather is just right, I can actually snowboard and water ski on the same day!"

The End of the Pipe

Ever wonder where all the used water goes? The water from showers, kitchens, washing machines, and, yes, toilets? How about the water that runs off from irrigated lawns or that is used to cool equipment in manufacturing plants?

In some rural areas, "gray" water goes to underground septic tanks. But in most towns and cities, gray water, called wastewater or sewage, goes to wastewater treatment plants.

Engineers who design and run these wastewater treatment plants are doing their communities a great service. Dirty, bacteria-laden sewage is treated so that it can be safely released into rivers, streams, or bays—odor free!

Ellen Richards founded a Women's Laboratory at MIT in 1875. The facility closed in 1883 when MIT began awarding more degrees to women, and men and women worked together in the same laboratory.

Background: Aerial view of a wastewater treatment plant.

Early Ecologist

There were no wastewater treatment plants for most of **ELLEN SWALLOW RICHARDS'S** life (1842–1911). Raw sewage flowed into city streets, down gutters, and into creeks and streams. As a result, diseases such as cholera were rampant. Drinking water supplies were foul, and fish caught in nearby waters were contaminated.

Ellen, who taught sanitary engineering at the Massachusetts Institute of Technology (MIT), implored her students to make cities cleaner through effective wastewater treatment. Graduates of her classes went on to design and build modern municipal sanitation facilities that prevented water pollution in the future.

This advocate for clean, healthy living is also considered the founder of the fields of ecology and home economics. In the late 1800s, the average home would be considered filthy by today's standards. Ellen applied principles of chemistry and sanitary engineering to remodel a house she called her "Center for Right Living." She installed upstairs windows that opened at the top or bottom to release stale air, and added indoor plants to provide more oxygen. She removed lead water pipes and redirected the home's waste system away from the drinking water well. All the principles of sanitation and ventilation she demonstrated are common practice today.

OVERCOMING OVERFLOWS

In some older communities, rainwater travels in the same pipes as sewage. As cities grow, these "combined sewer" systems become overwhelmed, and untreated sewage can flow into rivers and lakes.

Civil engineer **JUDY NITSCH** (b.1953) knows how to handle this problem. She teams with landscape architects, environmental engineers, and local officials to implement environment-saving plans using "ecohydrology." The team finds ways for rainfall to soak into the ground before it rolls down sidewalks and streets into the overtaxed sewer system.

"Ecohydrology mimics the way nature handled rainwater before development," explains Judy.

At Judith Nitsch Engineering, Inc., New England's largest woman-owned engineering firm, over 40 percent of the engineers are women. "Our women engineers are conscientiously contributing to make the environment better," says Judy.

JUDY NITSCH (LEFT) POSES INSIDE A MASSIVE PIPE USED FOR WATER SUPPLY.

MIGHTY MICROBES

Wastewater treatment, although a natural process, is complex. Modern wastewater treatment plants use small microorganisms, or "microbes," to break down sewage into harmless, non-toxic products.

Civil engineer **JOANN SILVERSTEIN** (b.1946) researches microbes that remove nitrogen, a major component of municipal wastewater. "There's no manual," she says of her experiments. JoAnn has also patented a treatment system for small communities. She was thrilled when a town in Oklahoma selected her system to treat contaminated groundwater over others because it was the most practical option.

ABOVE: BACTERIA SUCH AS THESE REMOVE CONTAMINANTS (IN THIS CASE, THE ELEMENT MANGANESE) FROM TREATED WASTEWATER BY TRANSFORMING CONTAMINANTS INTO DEPOSITS ON CELL WALLS (THE BROWN SPOTS ABOVE).

JOANN SILVERSTEIN'S LATEST PROJECT IS FINDING SPECIALIZED MICROBES TO TREAT ACID MINE DRAINAGE, A PROBLEM AT SOME FORMER ORE MINING SITES IN THE WESTERN U.S.

AWESOME OZONE

Chlorine has traditionally been the most common way to make drinking water safe and to make sewage clean enough to be discharged to nearby water bodies. To eliminate the need for chemicals, engineers are looking for ways to treat drinking water and wastewater without chlorine.

In Bossier City, Louisiana, **BARBARA E. FEATHERSTON** (b.1966) developed a water treatment program using ozone (a molecule containing three atoms of oxygen) instead of chlorine. "Ozone makes water safe for drinking and removes musty tastes and odors that can build up in some water supplies," says Barbara. Now Bossier City citizens enjoy water from the tap instead of buying bottled water or using personal water filters.

"As an engineer and mother, I feel proud whenever my family drives by a particular elevated water storage tank I've worked on. My kids always call out, 'That's Mommy's tank!'"

BARBARA FEATHERSTON INSPECTS A WASTEWATER TREATMENT PLANT IN SHREVEPORT, LOUISIANA, THAT USES ULTRAVIOLET LIGHT INSTEAD OF CHLORINE TO NEUTRALIZE BACTERIA AND BUGS IN SEWAGE.

In Tune With Nature

Every plant, every animal, the water in lakes and streams, even the air we breathe has a life of its own. The term "ecosystem" refers to the integrated collection of all these living forces.

Engineers work to find creative ways for people and the planet to live in harmony. They practice sustainable design to reduce the damaging effects our building and manufacturing activities have on the environment. And they help clean up pollution that was caused in the past.

Left: Half dome and the Merced river in Yosemite National Park. Above: Seashore in Hawaii.

Human beings are a natural force all their own. As communities grow, they affect the ecosystem. Human lives are precious, of course. We are learning to better balance our needs with the planet's greater good. Otherwise, the natural resources available to us now—clean water, air, and land—may not sustain us in the future.

Wetlands.

But what about when Mother Nature isn't so kind to humans in return? Natural disasters—such as floods, wildfires, tornadoes, and earthquakes—are a part of our life here on Earth. Engineers keep us safe, even when the earth flexes her muscles.

"The goal of life is living in agreement with nature."

Zeno
Greek Philosopher (490 BCE–425 BCE)

Back to Nature

What, truly, is more beautiful than nature? Which of humanity's creations rivals the rushing waterfalls in the Yosemite Valley? The Painted Desert of Arizona? The dewy marshland of the Everglades? The rugged Pacific coastline? Or the colorful trees of a New England fall? To the good fortune of Americans, many natural parks and preserves are maintained for all to enjoy.

But who protects the rest of the natural world? Women engineers do! They restore habitat for endangered species, rebuild wetlands damaged by previous development, and communicate that everyone must consider the environment in work and in play.

"Our work affects many different people: power companies, tribes, towns. The key is allowing everyone to feel heard while explaining what is technically important. I enjoy the challenge of communicating hydraulic engineering to non-engineers."

—Laurie Ebner

Above: Female bald eagle in flight.

Left: Everglades National Park, Whitewater Bay, Florida.

FRIEND OF THE FISH

Most species of salmon spawn—or lay their eggs—in fresh water like rivers and streams. The young fish migrate to salt water (bays or oceans) and grow up there. After they have reached maturity, the fish swim back upstream to fresh water to spawn.

As salmon journey downstream on the Lower Columbia River, they encounter several barriers, including the Bonneville Dam. There, many fish struggle to find a safe route to the ocean.

Civil engineer **LAURIE EBNER** (b.1957) is trying to save the endangered salmon. She studies the hydraulics—or the water flow—of the river, then shares the data with biologists in her U.S. Army Corps of Engineers work group.

"Together, we figure out the best way to help adult salmon reach upstream spawning grounds and juvenile salmon go downstream to the ocean," says Laurie. It's a complex situation! But the fish are a vital part of the ecosystem. Plus, the salmon is a symbol of dependability and renewal to local Native American tribes.

At Bonneville Dam, underwater screens guide the salmon away from the dam into a channel, placing them back in the river just downstream. "Measurements show more fish are surviving in this part of the river. But the real proof will come in three to five years when adult salmon return to spawn," says Laurie.

LAURIE EBNER, AN ENGINEER WORKING TO SAVE SALMON ON THE LOWER COLUMBIA RIVER, IS A WIFE, AND THE MOTHER OF A FIVE-YEAR-OLD.

Racing Concrete Canoes

In 2002, **AMY DESSNER'S** (b.1979) newfound degree in civil engineering landed her a job with the Natural Resources Conservation Service, an agency of the U.S. Department of Agriculture. Amy credits the local chapter of the American Society of Civil Engineers (ASCE) with helping her develop a career interest. While at the University of Wisconsin, Amy and her team entered ASCE's Concrete Canoe Competition. Her team designed and built a canoe using a thin layer of concrete for the shell. "And the canoe had to float," comments Amy, whose team won the boat race, but not the overall competition. Still, Amy prizes the experience she gained from the contest.

THE EVERGLADES: RIVER OF GRASS

A hundred years ago, the Florida Everglades were an untouched and tremendously dynamic ecosystem. The lush 4,500-square-mile marsh, described as a "moving river of grass," teemed with alligators, cougars, exotic birds, and hundreds of other species.

As people flocked to Southern Florida, though, the Everglades were drained dry to provide flood protection and drinking water to communities that sprang up.

Some women engineers were among the early activists to decry this major alteration of the natural world. Today, many engineers are restoring natural order to the Everglades.

SENIOR NATURE LOVER

At age 80, electrochemical engineer **MARTHA ELSMAN MUNZER** (1899–1999) moved to Florida where, as a member of Friends of the Everglades, she energetically opposed uncontrolled growth and urban sprawl. "When we lose touch with nature, we lose touch with ourselves," said Martha when asked why her crusade was important.

Her activism was no surprise to friends and family. Martha's love of nature began in Riverdale, New York, where she founded one of the nation's first environmental education programs, the Riverdale Outdoor Laboratories, in 1959.

After winning a zoning fight to preserve open space in Riverdale, a fellow environmentalist suggested that Martha become the director of the newly-formed Wave Hill Environmental Science Center, where children are still learning about nature's seasonal changes, sensitive gardening, and composting.

CONSERVATIONIST MARTHA MUNZER WROTE 11 BOOKS ON ECOLOGY AND CITY PLANNING. PERHAPS A GREATER ACCOMPLISHMENT TO SOME, SHE SWAM EVERY DAY IN AN UNHEATED POOL—EVEN WELL INTO HER 90S!

THE UNIQUE PLANTS AND ANIMALS OF THE FLORIDA EVERGLADES ARE AN AMERICAN TREASURE.

THE WATER RETURNS

The U.S. Army Corps of Engineers and South Florida Water Management District, the region's drinking water utility, are partners in restoring the Everglades as much as possible.

Army Corps civil engineer **CHERYL PHANSTIEL ULRICH** (b.1961), a mother of two, is a top manager in the Everglades Restoration Program. Cheryl and her team of environmentalists and engineers hope to recreate the Everglades' famed "river of grass" so native animals can return and the marsh will come back to life.

"We're trying to mimic water flow once natural in the area by controlling the quantity, quality, timing, and distribution of water there today," says Cheryl.

Another important part of her work is conferring with regulatory agencies, farmers, Native American tribes, and other special interest groups. "I help them understand that the restoration program will address their many, sometimes differing, concerns," says Cheryl.

"I stay hopeful that we will reach our goals because the goals are about the greater good of creating a sustainable South Florida, as well as the nation and the world."

—CHERYL ULRICH

CHERYL ULRICH STANDS IN A LARGE PIPE USED FOR THE WORLD'S LARGEST-SCALE ENVIRONMENTAL RESTORATION PROJECT, THE EVERGLADES RESTORATION PROGRAM.

Slip Sliding Away

Parts of Louisiana are literally disappearing. As much as 35 miles of the state's marshy coast erode away each year. And it's land no one wants to lose.

Louisiana's wetlands are a habitat for birds, fish, even certain butterflies. They are a source of income from commercial fishing and tourism. Wetlands keep Louisiana safe, too, by protecting inland areas from storms like Hurricane Katrina, and regulating water quality.

How did the problem start? Nature itself causes erosion. But activities to limit flooding, improve shipping channels, and aid agriculture have contributed as well.

Engineers are finding creative ways to save the Louisiana coast without harming communities or eliminating local jobs. One idea is to mimic natural water circulation so that natural sediments are returned to the coast. Another plan is to create marshes in some river basins and restore shorelines in other areas.

A Cleaner Environment

Humans have not always been the best caretakers of the earth's resources and natural habitats. During the early days of industrial growth in the United States, industries adopted an "out of sight, out of mind" policy for their chemical wastes. Companies dumped their toxic trash in vacant lots, drainage canals, or nearby lakes and streams.

These thoughtless actions have come back to haunt us. In the late 1960s, the Cuyahoga River, a tributary of Lake Erie, once contained so much debris and oil slicks that portions of the river caught fire when touched off by bridge workers' welding torches!

A sweeping movement towards environmental cleanup began with the tragic story of Love Canal. Women engineers have been at the forefront of these activities. Many see their work as a way to help hundreds, even thousands, of people with a single project.

"No man in the room could look at me and ask, 'What's she doing here?' because they knew I was an engineer. I was listened to."

—Rita Meyninger

Toxic contamination can affect wildlife as well as humans. For instance, birds in the area of an oil spill may die from direct contact with the oil or when oil vapors are inhaled. Other birds initially surviving the spill may die later when pollution destroys food sources in their habitat.

Stirring a Nation: Love Canal

A peaceful suburb built in the 1950s around Love Canal in Niagara Falls, New York became the nation's first publicized health and environmental hotspot. Starting in the mid-1970s, residents realized birth defects and slow growth rates among the area's children plus miscarriages and respiratory cancers among the adults were no coincidence.

They saw the oily scum and ooze entering their basements and yards. They saw their children coming home from play with burning faces and hands. They knew there had to be a connection.

By the time state health officials responded to their pleas, the people of Love Canal realized the full horror of their situation. They were living on top of a toxic time-bomb: Hooker Electrochemical Company had dumped chemicals, including dioxin, on land where homes and a school were then built on top.

In 1980, President Carter declared the area a health emergency. He appointed veteran civil engineer **RITA MEYNINGER** (b.1925), then a Federal Emergency Management Agency (FEMA) regional director, to coordinate evacuations of families living there.

"Right away, I realized the environmental and health studies at Love Canal would take years," says Rita. Some evacuated families were living in hotels. Other families still lived in the neighborhood with only a fence separating them from the toxic dump. "It was a lose-lose situation," says Rita.

She coordinated a huge FEMA team and convinced lawmakers to make all families in the study area eligible for immediate relocation to new neighborhoods. Congress loaned funds enabling everyone to leave quickly.

"I count my work at Love Canal as the most significant in my career because I helped get residents to new, safe lives," says Rita. "That's what civil engineering is about: public health and public safety."

The situation at Love Canal became a catalyst for the EPA's Superfund Program to locate, investigate, and clean up other hazardous waste sites around the nation.

Early in her career, Rita Meyninger (shown here in 1950) was in charge of 40 inspectors while monitoring quality control for a portion of the New Jersey Turnpike construction project.

Above: A sign warns of hazardous waste contamination at a school in Love Canal.

Right: The Hooker chemical plant dumped toxic waste in the Love Canal neighborhood from 1940 until 1950.

Dianne Dorland learned about the chemicals used in manufacturing as a researcher for Union Carbide and as a process engineer for DuPont.

Pollution P.I.

Even after wastewater is treated, trace amounts of mercury can remain that might pollute lakes and streams. Fish can absorb the mercury. If humans eat the fish, the mercury can affect their neurological systems.

In Minnesota, land of ten thousand lakes, mercury is the last thing people want in their water. It was there that chemical engineer **DIANNE DORLAND** (b.1947) became a pollution private investigator (P.I.).

The Western Lake Superior Sanitary District plant in Duluth, Minnesota, found mercury in its incoming sewage and in its treated wastewater and treatment by-products. Bad news! Dianne was called on to find the source of mercury.

"Our team tracked the mercury down to wastewater from a nearby pulp and paper mill. But the mill didn't use mercury," says Dianne. So where was the mercury coming from? She and the paper mill looked for suspects.

DIANE DORLAND became the first woman president of the American Institute of Chemical Engineers in 2003.

"We found that mercury was an impurity in a chemical used during the bleaching cycle. Curiously enough, the plant had just changed its bleaching process to reduce chlorine use—only to unintentionally introduce mercury," explains Dianne. The mill worked with its suppliers to eliminate the mercury sources. And the case was closed!

Harbor Hotspot

New Bedford Harbor in Massachusetts had a charming history as a whaling town, until it was declared a Superfund site in 1982. The harbor, considered the nation's most contaminated site, held enough PCB (polychlorinated biphenyls) and heavy metal sediments to fill 75 football fields, each three feet deep! Today, we know that PCBs can damage the liver, immune and reproductive systems, and even cause cancer.

Early in the Superfund program, civil and environmental engineer **YEE CHO** led a 45-member task force responsible for cleaning up the New Bedford Harbor and Landfill. Today, Yee is the founder and president of CDW Consultants, Inc., a firm that cleans up sites called "brownfields," urban sites slated for reuse and redevelopment.

Yee Cho uses surveying equipment to locate ideal spots for monitoring wells. Far right: Yee and Kathleen Campbell oversee drilling of monitoring wells to collect groundwater and soil samples. Samples will be tested for the presence of hazardous materials.

Waste Watcher

Careful handling and disposal of depleted nuclear fuel is critical because the waste is radioactive. Radioactivity can be harmful to people and the environment. While at the Nuclear Regulatory Commission, nuclear engineer **PATRICIA L. ENG** (b.1955) worked with the Environmental Protection Agency, other nuclear engineers, scientists, and the public to write policies regulating the use and disposal of radioactive materials.

In particular, she updated guidelines for transporting nuclear waste by truck on the nation's highways. "The new rules reflected the increase in private cars and driving speeds on U.S. highways in the late 20th century," says Patricia. She also considered the country's growing population, especially in communities near interstates.

Patricia traveled to towns near proposed nuclear waste storage sites to meet with the public. "I answered their questions and worked hard to ease their concerns about the nuclear waste coming near their neighborhoods," she says.

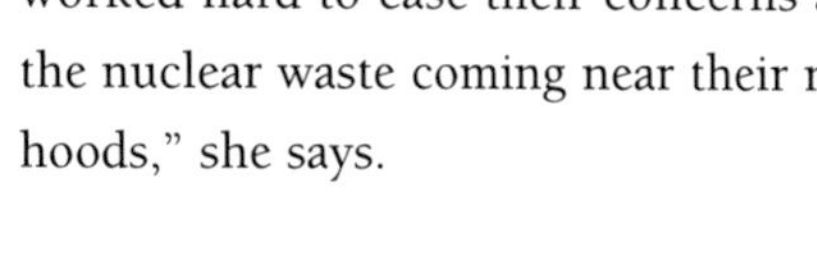

Patricia Eng became interested in science as a girl taking dance classes. Her dance teacher suggested she study physics when Patricia complained she couldn't do a triple pirouette. By the time Patricia was out of high school, she enjoyed all sciences and learning about the way things work. And, yes, she did finally do that pirouette!

"Don't assume you have to be a genius to study engineering. I'm a nuclear engineer, and I've locked my keys in my car!"

—Patricia Eng

Diana Bendz has made the world a better place not only by reducing the number of computers in the trash but by raising three children who, today, are all working scientists.

Computer Clean-up

As computer technology rapidly improves, more people regularly replace their old computers with better, faster models. But what happens to all those cast-off computers? Most wind up in landfills. Besides taking up valuable landfill space, chemicals inside the computers can leach out and damage the environment.

IBM responded to this growing concern. As director of environmentally conscious products for IBM, **DIANA KNIGHT BENDZ** (b.1946) developed a plan to design and manufacture computers with fewer negative impacts on the environment.

Soon, other manufacturers joined IBM in this pollution prevention task. As a result, today's new computers incorporate more recycled and non-hazardous materials. "Packaging has been minimized, and computers are more easily upgraded to lessen the need for new purchases," says Diana. Natural resources are further preserved by energy-saving "sleep" features on many models.

Our Cities and Towns

All engineering serves people and the public. However, engineers in "public works" are true public servants. They address the day-to-day needs of people in big cities, suburbs, and rural communities.

Public works engineers oversee the construction and repair of city streets, parks, and public buildings. They conduct citywide beautification projects. They make sure new developments are constructed properly. When a community is struck with severe weather or an unexpected event, they work with police and firefighters to safeguard citizens and public property.

At the same time, these engineers must work within a budget. They are, after all, stewards of the public's trust, and managers of a portion of the public's taxes!

Sustaining Cities

"Sustainability" is a hot topic for engineers in today's cities and towns. Sustainability refers to a broad level of environmental stewardship that maintains and enriches both natural and man-made resources. Local governments promote sustainability by:

- Reducing waste and energy use by recycling or reusing materials.
- Promoting environmentally safe disposal options.
- Complying with or exceeding environmental and health and safety standards.

FDR Drive, New York City.

The Big Apple Shines

Among civil engineer **GLADYS TAPMAN'S** (1912–1998) more "illuminating" projects for the bustling borough of Manhattan, was the placement design for every light along East River Drive, now called FDR Drive, from 49th to 99th Street.

The project was one of many that supported Manhattan's eastward expansion in the 1930s. A few years earlier, as part of the same civic growth efforts, Gladys estimated construction costs and materials needed to build an air-conditioned sanitation dump at 91st Street and the East River.

After graduating from Cornell University, Gladys Tapman's first job was estimating the quantity of steel and concrete needed to build a lock and dam in Quincy, Illinois, for flood protection along the Mississippi River.

S'NO KIDDING

Snow is no stranger to Erie County, home to Buffalo, New York. But a late-December snowstorm in 2001 left seven feet of snow over a large part of the county. Seven feet is as high as the typical front door to a house! The weight of the snow caused 280 buildings to collapse.

Civil engineer **MARIA LEHMAN** (b.1960), then the county's commissioner of public works, worked up to 19 hours a day pulling together a relief team of police, firefighters, emergency workers, engineers, and architects. "I used every bit of knowledge I had gotten from everywhere, and I did it in three days," she says.

That's right. Maria was part of the leadership team that dug the County—and its one million residents—out of the snow and helped them get back to business-as-usual in only three days! "And all while my own three athletic sons, then aged 13, 15, and 19, were cooped up at home due to travel restrictions," she adds.

Maria applied engineering principles to explain how homeowners, businesses, and schools could check their roofs for possible damage. She supervised the inspection of public buildings so workers would return to safe spaces.

AS PUBLIC WORKS COMMISSIONER, MARIA LEHMAN OVERSAW ABOUT 450 CONSTRUCTION AND IMPROVEMENT PROJECTS A YEAR AS DIVERSE AS A NEW "CSI STYLE" FORENSIC LABORATORY AND THE BUFFALO ZOO'S NEW SEA OTTER AND SEA LION EXHIBIT. ABOVE: MARIA OFTEN POSED FOR TELEVISION AND NEWSPAPER "PHOTO OPS" WHEN HER PROJECTS MADE THE NEWS. "MY ROLE WAS VERY VISIBLE," SHE SAYS.

LEFT: DECEMBER 2001'S ACCUMULATED SNOWFALL WAS THE MOST EVER RECORDED IN BUFFALO, NEW YORK'S, HISTORY.

"It is my responsibility as a professional engineer and a public works commissioner to uphold the safety of the public."

—MARIA LEHMAN

Iraq Improved

Over 17 million people in Iraq—about 70 percent of the population—live in urban areas lacking the most basic of services. Years of conflict, plus a lack of investment and maintenance during Saddam Hussein's regime, have left about 40 percent of city dwellers with poor access to clean drinking water.

Wastewater from millions of citizens flows untreated to the Tigris River or stands stagnant near schools, clinics, and markets because of inoperable sewage treatment plants. The health and environmental situation is unimaginable to people in developed nations!

Fortunately, architectural engineer **NESREEN M. SIDEEK-BARWARI** (b.1967) is on board as Iraq's minister of municipalities and public works. "I oversee the entire country's water supply, urban roads, and sanitation services," says Nesreen.

She is well-prepared for the challenge. Nesreen and her family, who are Kurds, were political prisoners who fled to Turkey with over a half a million other Kurdish refugees after the 1991 Gulf War. She served as the minister of reconstruction and development in the Kurdistan regional government to make sure her people were provided with housing, water, sanitation services, and access to schools and health-care facilities.

In Iraq, Nesreen's dedication to improving lives and limiting environmental damage is unwavering. She conducts her critical work amidst continuing unrest—despite a 2004 attempt on her life that left her driver and two bodyguards dead.

"I have prioritized 80 public works projects," explains Nesreen. "When they are complete, two major urban wastewater treatment plants will be operational, and drinking water will be delivered to 75 percent of Iraqis, the highest percentage in the nation's history."

Above: Nesreen Sideek-Barwari is working on a public works plan that gives local control to 266 municipalities so individual leaders can address their constituents' specific needs.

Left: Serving in a public crisis is nothing new to Nesreen Sideek-Barwari, the first woman to head a UN Field office (Iraq, 1997). She is pictured here in the Barwari area on the Iraq/Turkey border, her birthplace.

Flash Floods in Vegas?

Can the desert flood? Even though Las Vegas gets just over four inches of rain a year, thunderstorms in nearby mountains and the area's broad valley floor create the perfect conditions for flash floods. These unexpected floods have damaged buildings, crushed cars, made streets impassable, and taken lives.

As chief engineer/general manager of the Clark County Regional Flood Control District in the late 1980s and early 1990s, civil engineer **VIRGINIA VALENTINE** (b.1956) advocated for better flood control in Las Vegas. Tax dollars were needed, but it was a hard sell.

"Long time residents knew the dangers of flash floods," says Virginia. "But newcomers, pouring into Las Vegas at a rate of 5,000 each month, found the idea hard to imagine."

Once funding for Virginia's program was approved, construction began on new or improved flood channels, storm drain piping, and flood detention basins—each component designed to collect flood water before it can rush through the valley in a wave of destruction.

A storm in 1999 proved Virginia right. "Damage to Las Vegas was limited, even though construction of all the upgrades had not been completed," she says. Virginia later became the city manager of Las Vegas. She is now assistant county manager of Clark County.

The Red Rock Detention Basin, shown in the photo above, diminishes the deadly impact of flash floods by slowing flood water flow from as high as 75,000 gallons per second to just a few thousand gallons per second.

In a Flash

A flash flood can occur when a lot of rain falls in a short period of time. Within hours—literally in a "flash"—large quantities of water can bombard an area.

Flash flooding is common in valleys near mountain ranges, especially when alternate routes for the rain water are backed up with dirt and rocks or when flow channels are coated with minerals that cannot absorb some of the rain.

Mother Nature's Fury

Mother Nature can be serene and nurturing. Yet her repertoire also includes devastating floods, jolting earthquakes, unpredictable tornadoes, and sweeping hurricanes. Each can bring billions of dollars in property damage and the irreplaceable loss of human lives.

It doesn't take a once-in-a-lifetime event to cause injury. In the case of a flood, just six inches of fast-moving water in the street is enough to knock a person off her feet! Most cars will float away in just two feet of floodwater.

While natural disasters can't be prevented, many women engineers help minimize the destruction.

"Engineering is a great way to serve humanity, to make the world a better place, to improve quality of life—especially for those less fortunate—and to help preserve the planet's environment."

—Margaret Petersen

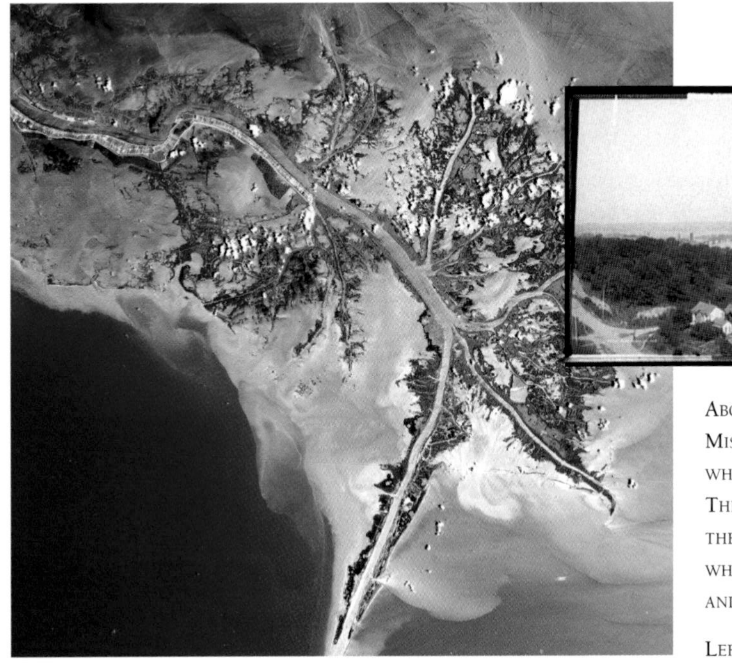

Above: The three vintage photos, circa 1900, show the Mississippi River flowing through Rock Island, Illinois, where hydraulic engineer Margaret Petersen grew up. This region is close to the "headwaters," or source, of the Mississippi. Flooding typically occurs downstream, where the river is its mightiest, in Missouri, Mississippi, and Louisiana.

Left: The mouth of the Mississippi River via satellite.

TAMING OLD MAN RIVER

Since the late 1800s, the U.S. Army Corps of Engineers has worked to control Mississippi River flooding. With rainwater from 31 states and Canadian provinces feeding the river, that's quite a job. The Corps has built levees, dams, and other flood control works to protect riverside development from deadly floods.

Growing up in Rock Island, Illinois, hydraulic engineer **MARGARET S. PETERSEN** (b.1920) learned how possessive the Mississippi River could be. The first floor of her great-aunt's house flooded almost every spring. "Over a 20-year period, the 15 steps leading from the porch to the riverside had been reduced to three," says Margaret. "Silty deposits had covered 12 concrete steps!"

After she graduated from college, The Corps hired Margaret for its flood control mission. Over 50 years ago, she helped build the Mississippi Basin Model, which helped engineers understand the river's response to floods and the operation of flood control works. This "scale model" covered 210 acres—a lot bigger than most models! Not only did the basin model help engineers forecast floods, it helped optimize communities' actions, such as sandbagging or evacuation, during floods.

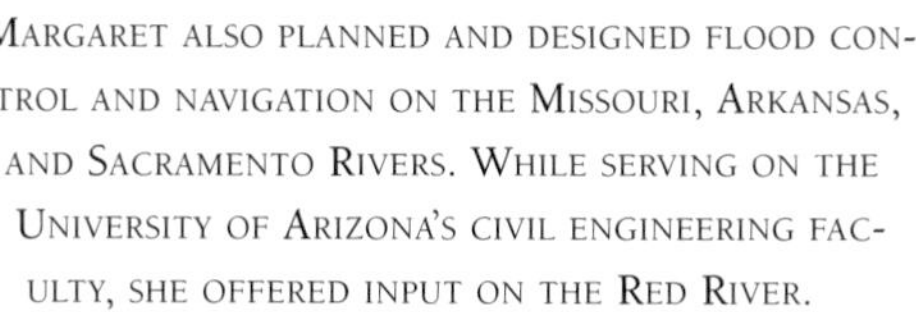
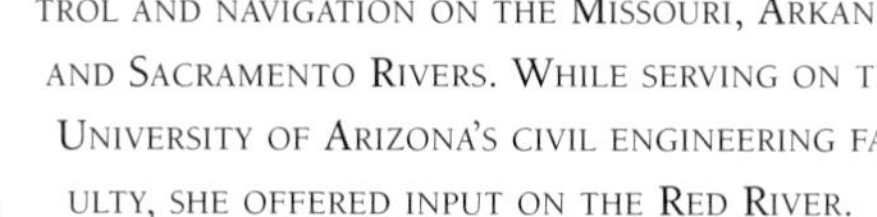

MARGARET ALSO PLANNED AND DESIGNED FLOOD CONTROL AND NAVIGATION ON THE MISSOURI, ARKANSAS, AND SACRAMENTO RIVERS. WHILE SERVING ON THE UNIVERSITY OF ARIZONA'S CIVIL ENGINEERING FACULTY, SHE OFFERED INPUT ON THE RED RIVER.

SHOCK WAVE SCIENCE

When the Earth quakes or bombs fall, the ground vibrates and the potential for damage spreads, or "propagates," in all directions. Much of the research on underground wave propagation is the work of civil engineer **ALVA MATTHEWS SOLOMON**.

Alva's study on the effects of bomb blasts and shock waves helped engineers determine the amount of concrete and steel needed to build strong missile silos during the Cold War. Her findings were also applied to earthquakes, which also send shock waves through the ground.

After Alva retired in 1983 and the Cold War ended, she thought other engineers might not find her work useful for new challenges. "It turns out I was wrong about that," says Alva. "Engineers use these theories today in designing buildings to withstand terrorist blasts."

By accurately sizing the expected shock wave, buildings can be fortified with additional materials or barricades. And new buildings can be located and sized to minimize damage.

AS A CHILD, ALVA SOLOMON LOVED THE DRAWINGS HER FATHER, AN ENGINEER, BROUGHT HOME. SHE DECIDED TO BECOME AN ENGINEER HERSELF. "AND NO ONE COULD TELL ME I COULDN'T DO IT," SHE SAYS. IN SPITE OF HER SERIOUS WORK IN SHOCK WAVE THEORY, ALVA (ABOVE) POSED FOR A 1950S FASHION SPREAD FEATURING WOMEN ENGINEERS. AT LEFT, ALVA TODAY.

"Being the only woman wasn't the greatest challenge. The real challenge was keeping up with rapidly advancing and challenging technology."

—ALVA SOLOMON

Women on the Go

For a long time, people relied on their own bodies and their own power to get around. Domesticated animals—horses, camels, oxen, and donkeys—were soon drafted into service, but it wasn't until machine power was harnessed that people really started to move!

Mobility means, simply, the ability of a person and their goods to travel from one place to another. For early humans, "mobility" meant survival. Today, mobility means freedom: freedom to explore, freedom to expand, freedom to improve our quality of life.

Above: Leonard P. Zakim Bunker Hill Bridge, Boston, Massachusetts. Left: Tokyo highway.

Today's transportation network is vast—and incredibly complex. There are cars and trucks and roads and bridges. There are trains and rails. There are airplanes and airports. There are ships and ports and harbors. And engineers had a big hand in building them all!

Top: View of the container terminal in the harbor in Hamburg, Germany.

Above: Three Eurostars in Waterloo Station, London.

WOMEN WITH WHEELS

Today, over three million vehicles are designed, engineered, and manufactured each year in the U.S.—and over 221 million cars, trucks, and SUVs are in operation. These vehicles provide "wheels" to 187 million registered drivers and their passengers.

Over the course of the century, the automotive industry has made revolutionary improvements in the comfort, safety, and performance of automobiles—many things that we take for granted today.

Behind each advance are major engineering challenges. How can vehicles be designed to reduce vehicular injuries or deaths? And how can vehicles be designed to reflect the unique needs and preferences of women? Women engineers have the answers.

In the U.S., on average . . .

- Each person in the U.S. takes 4 trips per day or 1,483 trips per year.
- Each person in the U.S. travels around 40 miles per day or 14,524 miles per year.
- Forty tons of goods are shipped, per person, each year.
- Each family spends around 18 percent of their household budgets on transportation, the second largest expenditure after housing.
- Fourteen million people are employed in transportation-related jobs.

The Windstar Moms

In 1999, more than 30 moms, 20 with children under age three, celebrated the arrival of the latest addition to their family—the Ford Windstar. As key members and engineers with the product development team, these "Windstar Moms" were involved in safety, ergonomics, electrical and fuel systems, product design and engineering, and climate control.

Interesting and innovative things happen when women design cars for themselves and their families! For example, the Windstar was the first vehicle to have "sleeping baby lights." The driver can open any door without shining overhead dome lights into a sleeping child's eyes. This feature has since been copied on other vehicles. Other family-friendly features included a larger gas tank to reduce the number of trips to the gas station, and added safety features that gave the Windstar a five-star crash safety rating.

Making a Difference at Ford

Cars have become much more safe over the past decade. Still, around 43,000 people die in car accidents every year. Mechanical engineer **SUSAN M. CISCHKE** is responsible for making Ford Motor Company's cars as safe, clean, and economical as possible.

As vice president of environmental and safety engineering at Ford, Sue is one of the highest-ranking and most influential woman engineers in the automotive industry. "I'm fortunate to see the direct effects of my work," she says, "in terms of lives saved and injury averted."

Sue is always looking to the future to see what Ford can do to improve the safety of its vehicles. She points to airbags and roll stability control as two active systems that have saved lives.

But, understanding that the vast majority of accidents are caused by driver error, Sue has also spearheaded the Driving Skills for Life program, which teaches safe driving techniques to teenagers nationwide.

One of mechanical engineer **ANNE STEVENS'S** first memories—at the age of four years old—was watching a stock car race on a dirt track at the county fair. "They absolutely fascinated me," Anne says. "Cars and motors are all about style, speed, adrenaline, passion, emotion. I love the industry."

Anne oversees all of Ford's business operations—purchasing, finance, sales, and marketing —in Canada, Mexico, and South America. "I spend a lot of my time focused on teams and relationships," she says.

Anne works on the "up, down, over, and out" philosophy of team building. "'Up' is the relationship with my boss," she explains. "Down" are the people that work for Anne. She notes that people will do almost anything for you if they know how it fits into the big picture and how it contributes to success.

"Over" are Anne's peers within Ford. She says that you can't even think about designing a car without consulting with the manufacturing engineers. The team has to develop a mutual understanding of what's required to build a new vehicle.

And finally, "out" means that Anne can't just focus on the internal operations at Ford. The world is changing fast, and she has to understand what might affect the automotive industry worldwide.

Not only is Anne Stevens Ford Motor Company's first female group vice president, she also races cars a couple of times a year. "I thrive on the speed and adrenaline," she says. Anne's daughter is a chemical engineer.

In addition to her responsibility for vehicle safety at Ford, Sue Cischke and her staff work to make cars more environmentally sustainable. Sue is proud of Ford's efforts to reduce the world's dependence on petroleum, which include this prototype hydrogen fuel cell electric vehicle.

A CAR FOR YOU

Automaker Volvo Car Corporation believes that its most demanding premium customer is the independent, professional woman. In the U.S., 54 percent of all Volvo buyers are women. That's why biomechanical engineer **CAMILLA PALMERTZ** (b.1967) and a cross-functional group of women presented a novel idea to the president of Volvo: develop a car by women, for everyone.

Volvo's president enthusiastically agreed and gave Camilla and joint project manager, mechanical engineer **EVA-LISA ANDERSSON** (b.1958), and other top women, free reign to develop a concept car with performance, prestige, and style. But it has more: smart storage solutions, good visibility, and minimal maintenance. It's also a car easy to get in and out of, a car that you can personalize, and a car that's easy to park. "A car is a very technical product," Eva-Lisa says. "Still, your buy is based on emotions."

THE FEMALE PERSPECTIVE WAS AT THE CORE OF VOLVO'S YOUR CONCEPT CAR (YCC) FROM THE BEGINNING. THE IDEA FOR AN ALL-WOMAN TEAM MAKING ALL THE DECISIONS AROSE IN THE FALL OF 2001. THE CONCEPT CAR MADE ITS NORTH AMERICAN DEBUT IN APRIL 2004 AT THE NEW YORK AUTO SHOW.

JOINT PROJECT MANAGERS CAMILLA PALMERTZ (LEFT) AND EVA-LISA ANDERSSON (RIGHT) BELIEVE THAT A CAR SHOULD MAKE LIFE EASIER, NOT MORE COMPLICATED. AS CAMILLA NOTES, "OUR AIM IS THAT YOU SHOULD FEEL GREAT IN THIS CAR."

VOLVO'S YOUR CONCEPT CAR WAS A PROJECT CREATED, DEVELOPED, AND MANAGED BY WOMEN INCLUDING (FROM LEFT TO RIGHT): MARIA WIDELL CHRISTIANSEN, EVA-LISA ANDERSSON, ELNA HOLMBERG, MARIA UGGLA, CAMILLA PALMERTZ, CYNTHIA CHARWICK, ANNA ROSÉN, LENA EKELUND, TATIANA BUTOVITSCH TEMM.

SHE PUT A "CAP" ON SMOG

By the early 1960s, there were over 74 million vehicles in the U.S.—by comparison, today there are over 253 million—spewing out carbon monoxide and other hydrocarbons. Smog, a term coined in 1905 as a combination of "smoke" and "fog," causes lung ailments and cancer.

Other engineers looked toward mufflers or afterburners to solve the problem of smog. But chemical engineer **VIRGINIA SINK** (1915–1986) had a new idea: control the exhaust by controlling the amount of gas burned.

She then came up with one of the first anti-smog devices: She called it the CAP, for "Clean Air Package." Virginia and her fellow engineers put the CAP on 500 test cars and reduced emissions to below California's standards at the time.

Virginia was personally recruited by Walter P. Chrysler in 1936 and became the first woman engineer at Chrysler. Among other things, she developed new auto surface coatings, high-speed corrosion tests, and a gadget that simultaneously measures acceleration, deceleration, cruising, and idling.

In 1965, Virginia stated, "The number and complexity of problems in industry is building up so that the number of engineers must continue to increase, and there is no reason why qualified women with engineering ability cannot find a logical place in the field." The same holds true today.

IN 1960, AUTOMOBILE MAKERS JOINED TO APPOINT A TECHNICAL PANEL OF CAR EXHAUST EXPERTS, LED BY VIRGINIA SINK, TO FIND A SOLUTION TO SMOG. VIRGINIA, SHOWN AT RIGHT WORKING WITH A CARBURETOR, CONTINUED HER WORK IN REDUCING VEHICLE EMISSIONS.

"Engineering is a way of thinking. You get the facts to form a broad sense of information from which you can draw and continue to learn."

—VIRGINIA SINK

Engineering, Race Car Style: A Weekend in the Life of Alba Colon

NASCAR (the National Association for Stock Car Auto Racing) is the fastest-growing spectator sport in America. Despite the fact that nearly half of the fans are women, on the track, NASCAR is a man's world . . . well, almost. Mechanical engineer Alba Colon (b.1968) manages the engineering services that General Motors provides to the Chevrolet teams that race in the NASCAR/Nextel Cup series.

NASCAR/Nextel sponsors 36 races a year, and Alba is at every one, right there with the action. Her job is to help her teams win! Here's a typical weekend in the life of Alba Colon.

Alba's weekend begins on Friday, a busy day of practice and qualifying sessions. There are race cars coming and going from the garage area to the race track. The crews hover over the engines talking about how to get the cars to go faster.

Alba's first stop is to talk with the NASCAR directors to go over some technical issues. All of the cars are inspected on Friday and Alba wants to know the results. She also wants to see the data. Are any of Chevrolet's competitor teams bending the rules?

It's Sunday, the day of the race. The drivers are strapped in. Cars are revving their engines. The flag waves and the cars are screaming around the track—reaching average speeds of 143 miles per hour! Alba watches the Chevrolet cars carefully.

Next stop: the crew chiefs. Alba makes the rounds, talking to all 17 of them and their crewmembers. Two weeks ago, five of Hendrick Motorsport's engines failed. Alba and her team of engineers figured out that an experimental part they had used failed. She didn't want that to happen again!

We won! NASCAR driver Tony Stewart gives Alba a big thank-you kiss after receiving the race trophy.

On Monday morning it's time to troubleshoot. Three Chevrolet cars had engine troubles yesterday and didn't complete the race. Why? Alba talks to the crews and then gets her engineers back in Detroit on the problem.

Liquid Fuel for our Mobile World

What would a day in the 21st century be like without gasoline to power cars, transit buses, and delivery trucks? What would happen to our international economy and tourism industry without petroleum products to fuel airplanes and ships? Engineers deliver liquid fuel to our mobile world.

Today, Decie Autin (left) manages deepwater oil and gas development projects in Nigeria. There, local tradespeople are learning underwater welding techniques, so her efforts enhance the country's economy and skills base. Marilyn Tears' (right) work on multinational teams has taken her to Korea, Singapore, Spain, Finland, Angola, and Chad.

Left: A major deep water drilling achievement, the Hoover/Diana is located in waters nearly one mile (4,800 feet) deep!

Black Gold

The Hoover/Diana is an oil and gas production platform 160 miles south of Galveston, Texas, in the Gulf of Mexico.

From 1998 to 2000, civil engineer **MARILYN TEARS** (b.1957) and chemical engineer **DECIE AUTIN** (b.1959) worked together to get the platform up and running. Once built, the platform was the largest and heaviest of its kind. Non-stop operations are designed to process 325 million cubic feet of gas and 100,000 barrels of crude oil every day—enough to fuel two million vehicles!

As operations superintendent, Marilyn developed the platform's operating team. She selected personnel, handled training, and managed the business unit for this $1.1 billion asset.

Decie, the start-up and technical manager, oversaw a long checklist of pre-start-up tasks and handled problems that threatened to throw the project off schedule.

Their hard work paid off, and the platform began production ahead of schedule. Marilyn and Decie agree that a combination of engineering and people skills were key in getting both equipment and work teams functioning smoothly.

You're Safe!

For chemical engineer **DEBORAH L. GRUBBE** (b.1955), watching out for worker safety is the ultimate level of commitment to professionalism. Deborah knows that when companies truly care about safety, workers feel secure and motivated. "A safe workplace is good for business," she says.

While at chemical company DuPont, Deborah led two large manufacturing organizations in making products safely. She also led a team of 700 engineers that worked all around the world to help teams be safer and more productive.

Deborah believes that it is important to make a strong commitment to one's profession. She also believes that her passion for safety and for excellence has been a factor in her career success.

Deborah is now vice president of group safety and industrial hygiene for BP. She not only directs programs to make sure workers are safe, she oversees efforts to keep customers, suppliers, and the people who live near plant sites safe as well.

Deborah Grubbe's work in health and safety draws on her broad experience in chemical engineering lab work, design and construction, and other engineering management roles.

Pipeline to Power

Keeping petroleum pipelines free of rust was **IVY M. PARKER'S** (1907–1985) specialty. She developed filters, selected the best corrosion inhibitors, and wrote maintenance plans to keep pipelines and storage tanks clean. Otherwise, they would have to be shut down to clear out the corrosion.

While at Plantation Pipe Line Company, Ivy was responsible for 2,400 miles of pipeline from Baton Rouge, Louisiana to Greensboro, North Carolina. Her efforts in beating rust allowed the company to double its gasoline flow to customers.

Ivy Parker was honored as the "First Lady of Petroleum" by professionals in the petroleum field.

Roberta Nichols's own car was a late-1950s model Mercedes 300 SL Gullwing. She achieved a top speed of 145 miles per hour in this car.

Fuel Diversity

It all started with water skiing on the Colorado River with friends. One thing led to another, and mechanical engineer **ROBERTA NICHOLS** (1931–2005) began racing her boat against her friends' boats. "The challenge was trying to get more horsepower out of the factory-built engine while making sure the engine wouldn't fall apart during a race," she says.

Roberta held the world drag boat racing record from 1966 to 1969: the fastest woman to go a quarter mile on water. After "seeing too many friends die," she says, Roberta turned her sights to racing on land.

At Ford Motor Company, Roberta was a renowned expert in alternative fuels. She developed cars that could run on ethanol, methanol, or gasoline. (Ethanol is often used in professional racecars because of its high-performance features.)

Roberta borrowed a friend's car, a 1929 Ford Roadster High-Boy with a Chrysler Hemi Engine running on ethanol, took it to the Bonneville Salt Flats in Utah, and reached a top speed of 190 miles per hour!

Women Love Cars!

Women love cars and women engineers love to make them. As electrical engineer **IRENE SHARPE** (b.1941) who specialized in automotive wiring says, "you would be out driving along, and there goes [a car] that has your parts on it. And that to me was the most exciting part of my engineering career: to be out there on the road and have those cars go whizzing by me . . ."

Women in the U.S.:

- Buy over half of the cars sold
- Influence around 80 percent of the car-buying decisions
- Buy 40 percent of all sport utility vehicles (SUVs)
- Spend 17 weeks on the new car buying process
- Rate safety as the most important aspect

CERAMICS FOR CLEAN AIR

Corning Incorporated makes the materials used in a bunch of products—one of which is the catalytic converter (which reduces air pollution from cars). Electrical engineer **LAURA J. MECHALKE** (b.1968) has worked for Corning since 1990 manufacturing lighter, more effective catalytic converters.

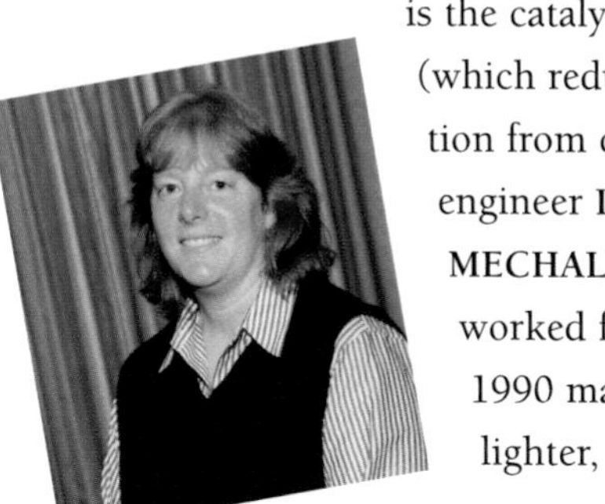

TODAY, LAURA MECHALKE IS IN CHARGE OF QUALITY FOR CORNING'S LATEST PRODUCT TO CLEAN THE AIR: THE CORNING DURATRAP® AT FILTER, WHICH WILL HELP DIESEL AUTOMOBILE MANUFACTURERS MEET INCREASINGLY STRINGENT PARTICULATE EMISSIONS REGULATIONS.

In her most challenging project, she streamlined the manufacturing of the ceramic core of catalytic converters. The upgrade required new equipment, so Laura also helped change the way workers did their jobs. Rather than have individual workers specializing in only one process step, Corning switched to a team approach.

Everyone on the team could do any step. Workers could flow to a temporary production bottleneck or rotate into another position. The workers themselves contributed ideas for these changes based on their hands-on plant experience.

"It's easy to change equipment. It's challenging to change the culture of a work force. What works is treating people with respect, listening to them, and valuing their opinions and experiences."

—LAURA MECHALKE

ULTRA CLEAN

Diesel engines have been around for over 100 years. The emissions (exhaust) from diesel buses, trucks, and cars have been getting cleaner and cleaner over the past 30 years. In 2007, new even more stringent emissions standards (ten times lower for particulates than soot!) will kick in.

In 2000, **RODICA BARANESCU** became the first woman president of the Society of Automotive Engineers.

"There's no doubt that International Truck and Engine will meet the new standards," says mechanical engineer **RODICA BARANESCU** (b.1940). "International has been developing emissions-compliant engines for over 25 years. All of our new school buses, for instance, already meet the standards for 'Green Diesel Technology,®' a design that significantly lowers the emissions and odor of diesel-powered buses and trucks."

For the last few years, Rodica has concentrated on the fuel, lubricants, and coolants for diesel engines. "To burn clean, the advanced technology engines need ultra-low-sulfur fuels. And the lubricants need to be compatible with the fuel," she explains. "This has been quite a challenge for the petroleum industry, but it's an important part of our nation's strategy for cleaner air."

RODICA BARANESCU ADVOCATES A "SYSTEMS APPROACH" TO ENGINEERING: "EVERYTHING IS PART OF A LARGER WHOLE. UNLESS YOU'RE AWARE OF THE OVERALL SCOPE, YOU'LL DEVELOP SOMETHING THAT DOESN'T MATCH WITH THE REST OF THE SYSTEM."

The Road More Travelled

How do you get from here to there? Let's see. In the U.S., you're likely to travel on part of the nation's four million miles of roads—the street outside your home, the two-lane highway, the interstate. Did you know that cars and trucks travelled 2,855,756,000,000,000 miles on those roads in 2002? That's nearly three quadrillion miles!

Roads and highways need to be convenient. They need to be safe. And in today's busy world, they need to handle a lot more traffic than they've ever seen before. All sorts of creative ideas are needed to keep those cars and trucks moving—which gives credence to the phrase, "a woman transportation engineer's work is never done."

Queens of the Road

It was 1956. Dwight D. Eisenhower was president. And cars were king: big cars with chrome bumpers and tail fins. Congress set aside $350 million to build an interstate highway system, linking major cities across the U.S.

With the opening of the Arroyo Seco Parkway in Pasadena in 1940, the California Department of Transportation (Caltrans) led the nation into the "freeway era." Now it was time to expand. Caltrans needed engineers and women stepped up to the challenge!

When civil engineer Ethyl Ann Hansen joined the California Department of Transportation (Caltrans) most of the women were either secretaries or draftspersons. Ann joined as an engineer fresh out of college and spent her entire career working to improve the transportation network in the San Francisco Bay area. She was the first woman to become a Caltrans deputy district director. After the Loma Prieta earthquake in 1989, she was responsible for keeping traffic flowing, even though a section of the Bay Bridge had collapsed and several major highways were closed.

Background: Freeway 110 and 105 interchange, aerial view, Los Angeles, California.

LA's San Diego/Santa Monica freeway interchange (I-405/I-10) is one of the busiest in the world, and a woman designed it! Civil engineer Marilyn Jorgenson Reece (1926–2004) used graceful loops and gentle curves to create the interchange, which also became one of the most photographed freeway interchanges in the world. One of Marilyn's daughters, Kirsten Stahl, is now an engineer with Caltrans, and her other daughter, Anne Bartolotti, chose a career on the "Information Superhighway."

Marilyn Reece and Carol Schumaker—shown here standing in front of the three-level Santa Monica-San Diego freeway interchange—were featured on the front page of the *Los Angeles Times* on April 6, 1964. Women engineers were such a novelty at the time that the headline read, "Freeway Builders Are Weekend Housewives: Highway Engineers Look Forward to Ordinary Suburban Chores Around Home."

Like many, Lois L. Cooper (b.1931) began her career with Caltrans in the heyday of highway building in Southern California, the first African-American woman hired in the engineering division. She participated on designs for many of the largest Los Angeles freeways: the Century Freeway, the San Diego Freeway, the Long Beach Freeway, the San Gabriel River Freeway, and the Riverside Freeway.

"On our roads, we try to design wide-angle or long, sweeping curves and gentle elevation of banking," civil engineer Carol Schumaker (b.1929) told an interviewer for *McCall's* magazine, May 1965. "Comfortable driving demands gradual grades to make the ride as smooth as possible." Carol was a chief engineer in the Caltrans route planning section and worked on the San Diego Freeway (I-405).

SHE HAS STREET SMARTS

Over 10,700 athletes from 197 countries—and two and a half million people—visited Atlanta during the Summer Olympic Games of 1996. Centennial Olympic Park was their gathering place.

To see the park today, it's hard to imagine that the area used to be a collection of rat-infested abandoned buildings and trash-strewn empty lots. Hard, also, to imagine how people used to get around that neighborhood, now that civil engineer **MARSHA ANDERSON BOMAR'S** (b.1952) traffic plan is in place.

Marsha recommended sweeping changes to the flow of traffic around the park and convinced the park designers that her concept was the right one. They agreed.

Today, a number of streets are completely closed, bus routes have been redirected, and better access to and from the nearby interstate highway is in place. "The traffic worked amazingly well during the Olympics," Marsha observes, "and it still works well today."

MARSHA ANDERSON BOMAR was the first woman president of the Institute of Transportation Engineers.

"Just goes to show you what can happen when you have street smarts," (which, by the way, is the name of the very successful transportation planning and engineering firm Marsha founded in 1990!)

ABOVE: THE 21-ACRE CENTENNIAL OLYMPIC PARK IN ATLANTA MADE ITS DEBUT IN 1996 AS THE WORLD'S GATHERING PLACE DURING THE CENTENNIAL OLYMPIC GAMES.

LEFT: TRANSPORTATION ENGINEER MARSHA BOMAR COMMENTS ABOUT HER CAREER, "ENGINEERING IS A PLACE WHERE YOU REALLY HAVE THE CHANCE TO MAKE A DIFFERENCE IN THE LIVES OF EVERYONE AROUND YOU—AND THE WORLD BEYOND. FIND THE NICHE THAT FITS YOUR SKILLS AND PERSONALITY!"

Construction in the Negative

Constructing "in the positive" means that the building occupies space that was formerly empty. Tunnels, shafts, and other underground spaces are built "in the negative," that is, carved out of something that already occupies space.

The problem with underground construction? Imperfections or unknowns lurking in the surrounding earth that may cause problems. These uncertainties run the gamut: a hidden layer of loose sand, unexpected faults, flowing groundwater.

PRISCILLA PROVOST NELSON (b.1949) has dedicated a good part of her career to trying to evaluate the uncertainty—or risk—of building tunnels, including the Buffalo Metro Rail, the Chicago Tunnel and Reservoir Plan, the Superconducting Super Collider, and others.

"As the world becomes more densely populated, there will only be more and more uses for underground space," Priscilla says. "The underground is certainly not the easiest place to work as an engineer, but it is the most satisfying I can imagine."

Priscilla Nelson is one of the first women to be offered membership in "The Moles," an elite group of leaders in the tunneling industry. She's also probably one of the first to see the inside of a tunnel boring machine (above), the monster that claws its way through the underground to create tunnels.

Left: Priscilla takes a ride on a prototype magnet for the Super-conducting Super Collider, a particle accelerator collider built in a tunnel near Dallas, Texas to study high-energy physics. Priscilla helped engineer the tunnel.

Priscilla Nelson worked with the planning team on the tunnels for the Washington Metro.

Building Bridges

A bridge can represent so many things: a way to get from one side to another, or a structure so beautiful that it becomes a signature landmark for its city or town. It can also represent the way people come together for a common purpose.

To build a bridge is an amazing feat! A huge structure weighing tons and tons has to be suspended over the empty space above a deep gorge or a wide river.

What we see when we cross a bridge is the steel and concrete—the graceful arches or elegant suspension cables. What we don't see, but what the bridge really represents, is the work of a team of thoughtful engineers and constructors who have come together to create a work of art.

The Bridge That Couldn't Be Built

The Brooklyn Bridge, which crosses the East River in New York City, was one of the most monumental engineering triumphs of its age. People thought it couldn't be done, but John Roebling had a solution: wire ropes that would suspend the bridge over the river. John died of tetanus before construction began, but his son Washington Roebling continued the project.

Three years into construction, Washington's health failed and his wife, **EMILY WARREN ROEBLING** (1843–1903) took over. For 11 years she made daily on-site inspections, dealt with contractors and materials suppliers, handled technical correspondence, and negotiated political issues.

When the Brooklyn Bridge opened in 1883, Emily rode in the carriage with U.S. President Chester Arthur. Congressman Abram S. Hewitt, speaking at the opening ceremony, praised Emily, saying, "One name, however, which may find no place in the official records, cannot be passed over here in silence . . . The name of Mrs. Emily Warren Roebling will thus be inseparably associated with all that is admirable in human nature, and with all that is wonderful in the constructive world of art."

Despite advice given to her when she was young that women didn't need a higher education, Emily Roebling studied mathematics and science. After her husband fell ill during construction of the Brooklyn Bridge, Emily began studies of her own on the engineering issues. She became so good at taking over for her husband that many suspected that Emily's was the great mind behind the great bridge!

Left: Brooklyn Bridge Poster, 1883. Background: Brooklyn Bridge.

SHE EXPLAINS "WHY"

A project as huge as the reconstruction of the Woodrow Wilson Bridge over the Potomac River between Maryland and Virginia is bound to generate some controversy. After all, every day nearly 200,000 vehicles cross the bridge. The roads leading up to the bridge pass through historic and large residential communities.

Civil engineer **NORINE WALKER'S** (b.1959) job is to communicate with the public about the project and listen to their concerns. Norine conducts public meetings, appears on local cable TV, takes people on tours, staffs a project hotline, manages a project website, and works with local students to incorporate the principles of bridge engineering into their curriculum.

Norine is an outgoing dynamo, even though she has severe rheumatoid arthritis—the nation's number one cause of disability. "Some days are better than others," she says, but adds, "I'm passionate about my life and how I've chosen to live it."

"WE'LL GET THROUGH THIS TOGETHER," NORINE WALKER EXPLAINS TO PEOPLE WHO MAY VOICE CONCERNS OVER THE WOODROW WILSON BRIDGE CONSTRUCTION. NORINE BELIEVES THAT WHEN PEOPLE UNDERSTAND WHAT'S HAPPENING AND WHY, THEY'RE MUCH MORE LIKELY TO APPROVE OF THE PROJECT.

THE NEW FOUR BEARS BRIDGE (ABOVE) IS EMBELLISHED WITH 14 MEDALLIONS THAT REPRESENT THE NATIVE AMERICAN CULTURE OF THE AREA. LINDA FIGG, SHOWN HERE WITH TEX HALL, PRESIDENT OF THE NATIONAL CONGRESS OF AMERICAN INDIANS, WORKED CLOSELY WITH A CULTURAL ADVISORY COMMITTEE TO ESTABLISH THE BRIDGE'S AESTHETIC ELEMENTS.

WHEN BRIDGES ARE ART

Bridges aren't just a way to get from one spot to the next! Rather, they serve as distinctive landmarks that express the spirit of the communities and environments they serve. "Each bridge has its own story," says **LINDA FIGG**, a 1981 civil engineering graduate.

The story of the Four Bears Bridge over Lake Sakakawea in North Dakota is inextricably linked to three Native American tribes—the Mandan, Hidatsa, and Arikara—that live on nearby Fort Berthold Reservation.

Using the FIGG Bridge Design Charette™ process pioneered by Linda and the FIGG team, members of the Three Affiliated Tribes were involved in creating the bridge's design story. Together, they created many of the bridge's distinctive features including bridge shapes, patterns cast into the pedestrian walkway, and the design of 14 four-foot-diameter medallions along the pedestrian sidewalk that represent tribal culture and history.

"I'm thrilled when people in a community become excited about their bridge," says Linda. "And I'm passionate about creating bridges as art."

LOCOMOTION

The steam locomotive was one of the most profound advances brought to society by the Industrial Revolution. The first railroads began to operate in the mid-1820s. Overnight, it seemed, farmers and factories were shipping their wares farther than they ever imagined.

The completion of the Transcontinental Railroad in 1869 was a turning point in U.S. history. The railroad tied the nation and its people together. Anything was possible! Millions of people traveled by trains and ships. Some trains, like the Orient Express, were elegant, opulent, glamorous. Others were hot, stuffy, and cramped.

Some say the golden era of travel by train is over. This may be true for long-distance travel, but millions of people still take the train. Subways, people-movers, and light rail have replaced the traditional railroad.

An estimated 14 million Americans ride public transportation (trains and buses) each weekday, and an additional 25 million use it on a less frequent but regular basis. In 2000, Americans took 9.4 billion trips using public transportation

ABOVE: NOTTING HILL GATE STATION, PART OF LONDON'S UNDERGROUND (OR TUBE) SYSTEM.

LEFT: BULLET TRAIN, JAPAN

WORKING ON THE RAILROAD

When **OLIVE W. DENNIS** (1885–1957) joined the Baltimore & Ohio (B&O) Railroad in 1920 with a civil engineering degree from Cornell, most train travel was hot, dirty, smelly, and uncomfortable. Olive joined as a draftswoman in the bridge engineering department.

It only took a year for the president of the B&O to recruit Olive for a special assignment. Knowing that half of the passengers were women, he wanted a woman with technical training to make a survey of the whole train system. Olive became an "engineer of service" and was to suggest ways to improve B&O's service.

Olive always stressed that her position was wholly an advisory one, that she never did anything but "suggest." But many of her suggestions became reality: reclining seats with footrests and headrest covers, a special window vent to allow fresh air into the coaches, and, eventually, fully air-conditioned trains.

Airplanes started to carry passengers in the late 1920s, and the railroads were keen to know the competition. Olive Dennis flew for the first time in 1928, and described the experience: "Every few minutes the plane would drop on an even keel, like a suddenly descending elevator, and the sinking sensation thus created began to concentrate at the pit of my untrained stomach, which just naturally resented it." She was airsick, in other words.

THINGS WE TAKE FOR GRANTED TODAY ON TRAINS—AS WELL AS ON BUSES AND AIRPLANES—WERE OLIVE DENNIS'S WORK: LARGE RESTROOMS FOR WOMEN WITH FREE PAPER TOWELS, LIQUID SOAP, AND PAPER DRINKING CUPS; OVERHEAD LIGHTS THAT COULD BE DIMMED AT NIGHT; INDIVIDUAL READING LIGHTS; AND SNACK COUNTERS IN EACH CAR.

"No matter how successful a business may seem to be, it can gain even greater success if it gives consideration to the woman's viewpoint. . . . Any organization wishing to sell its products or services to both must take into account the differences between them."

—OLIVE DENNIS

MARCH 1954 ADDRESS TO THE SOCIETY OF WOMEN ENGINEERS

People are Moving!

Clark County, Nevada, has nine automated people-movers—monorail trains—to help the public navigate all the action on the Las Vegas Strip. People-movers are a great way to get lots of people from one spot to another.

The Clark County Building Department hired **DIANE MORSE** (b.1957) to inspect construction of people-movers. "The Las Vegas Monorail was the first of its type to be built in the world, and we had to make sure it was safe," says Diane.

There were no guidelines on building these unique trains. "We were using amusement park ride building codes," she recalls.

Over the next four years, Diane became an expert. She wrote a county ordinance, still in use today, that lays down the rules on how companies can construct and operate people-movers in Las Vegas. Diane's work on standards now helps countries around the world build their own people-movers.

Diane Morse's experience with the CSX Railroad was quite helpful when she had to establish standards for people-movers in Las Vegas. She's now a renowned expert in the field.

Right: The Las Vegas Monorail was the first of its kind in the world. It glides above traffic at speeds reaching 50 miles per hour!

New York Transit Authority Chief Engineer Connie Crawford's favorite project took place when she helped rebuild the Brooklyn Bridge, originally engineered by Washington Roebling in the 19th century. Connie developed a computer model to replace all of the suspenders and cable stays that hold up the bridge, but no one could find the original design information for her to check her model. Through a tip, Connie uncovered Washington's diaries with his hand-written entries detailing daily developments based on conversations he had with his wife Emily. Washington became an invalid in the early stage of the bridge's construction, so Emily carried on management duties through daily discussions with Washington at home and with the construction foremen at the site. (See page 88 for the profile of Emily Warren Roebling.) With the original information gleaned from Washington's diaries, Connie was able to complete her task.

9/11 Underground

When two planes exploded into the World Trade Towers in New York City on September 11, 2001, massive devastation extended below ground as well as above. A 1,400-foot length of subway tunnel was crushed beneath one of the collapsed towers, and another 1,400 feet were seriously compromised by the catastrophe.

The New York City Transit Authority (NYCTA) took immediate action. At the time, civil engineer **COSEMA "CONNIE" CRAWFORD** (b.1956) was Deputy Chief Engineer of the NYCTA. "We had to get the subway up and running again fast so people could resume fairly 'normal' lives," recalls Connie.

Usually the design time on a project like this would have taken two to three years, "but we got it done in six weeks. Construction started immediately—24 hours a day, seven days a week." Within one amazing year, subway service was restored, four or five years ahead of a conventional schedule. The troubling gap on the lower Manhattan Subway map was filled.

ALL ABOARD!

"Whenever anyone visits the U.K., they come to London," says materials engineer **KAREN FERGUSON** (b.1965). "Increasing traffic congestion meant it was becoming more and more difficult for people to travel around the city. Something had to be done."

That "something" was an extraordinary undertaking: bringing all of London's transportation systems—roadways, traffic lights, cars, buses, the Underground, taxis, river services, and railways—into a new integrated organization, called Transport for London, or TfL.

Karen managed the change of ownership of the Underground—London's subway and one of Britain's world-renowned icons—from British government control to TfL.

How did Karen pull this off? "Project management by people management," Karen states. "I concentrated on using my 'soft' skills—leadership, motivation, and facilitating communication—which I supported using a simple but disciplined program management methodology."

Karen broke down the program into a series of small, well-defined tasks for people to deliver each week while she monitored the progress and controlled the overall strategy in the background. The beauty of her approach? "Each team member constantly achieved results. They remained focused, committed, and motivated."

ONE SENIOR ADVISOR FROM TRANSPORT FOR LONDON LIKENED KAREN FERGUSON'S PROGRAM MANAGEMENT SKILLS TO A "CONDUCTOR OF A GREAT ORCHESTRA." KAREN KEPT THE TRANSPORT FOR LONDON, LONDON UNDERGROUND, AND BRITISH GOVERNMENT TEAMS WORKING TOGETHER.

TAKE THE FIRST TRAIN OUT

There were a lot of naysayers in Dallas in the late 1980s and early 1990s, recalls civil engineer **FARIBA NATION** (b.1961) of DMJM Harris. "No one will ever ride DART rail. You simply won't get Texans out of their cars," they said. DART (the Dallas Area Rapid Transit) was the Lone Star state's first light rail transit system.

"DART exceeded its ridership projections the first year of operation, and it's been exceeding them ever since," Fariba notes with a smile.

FARIBA NATION IS A TOTALLY HANDS-ON ENGINEER. DURING THE START-UP OF DART, SHE EVEN WENT OUT WITH A STOPWATCH TO MEASURE THE TRAIN'S SPEED AND THE TIMING AT THE RAIL CROSSINGS

Fariba has spent nearly her entire career making DART light rail a reality. She originally worked on the 20-mile starter system in Dallas and then the 24-mile extension serving nearby cities.

Fariba and her team are responsible for putting new DART rail segments into operation. This means testing the signals and communications and traction power—millions of details that have to work right.

"My biggest reward is taking the first train out on a new line," says Fariba. "It sends chills down my spine to know that I gave the project my all and made it into something!"

Laden Ships, Lively Ports

The smell of salty sea breezes, the feel of the wind through their hair, their pride in knowing that the ships and ports they build are the best in the world—these are just a few of the pleasures that women engineers receive from their maritime careers.

More than two billion tons of freight are shipped through the 185 deep draft seaports in the U.S. every year! The leading cargo shipped through U.S. ports (imports and exports) includes petroleum, chemicals, coal, food, wood, and iron and steel. Tons and tons of consumer goods such as cars, clothes, and electronics are also shipped by sea.

People go through ports, too. Over 134 million passengers are transported by ferry and five million on cruise ships every year. That's a lot of comings and goings!

Ship Shape

After attending Bryn Mawr, **LYDIA GOULD WELD** (far right) graduated in 1903 from the Massachusetts Institute of Technology with a degree in naval architecture and marine engineering. She worked for the Newport News Shipbuilding and Dry Dock Company in Virginia making finished plans for government ships.

Full Speed Ahead

Northrop Grumman Newport News builds and maintains nuclear-powered aircraft carriers and submarines, the most sophisticated ships in the world. Mechanical engineer **ALMA MARTINEZ FALLON** (b.1958) began her career at Newport News during her sophomore year at college.

Alma's favorite assignment so far was working on the *USS John C. Stennis*, a nuclear-powered aircraft carrier—which is a large warship that carries airplanes. The *USS Stennis* has a long flat deck (1,092, feet to be exact—nearly as long as four football fields placed end-to-end!) for take-offs and landings.

Alma helped design the cooling systems for the propulsion plant—a critical element in making the ship move forward. "One of the coolest experiences I've had was going on sea-trials. I was responsible for making sure some of the ship's auxiliary systems in the engine room were sea-worthy and ship-shape."

Today, Alma serves as the hull structural construction superintendent for the CVN-21, one of a brand new class of aircraft carriers. "The CVN-21 is as long as the Empire State Building is tall," Alma notes. "I manage all of the steel fabrication: the steel beams, the exterior hull, and the bulkhead—everything that will make the boat float."

Alma Martinez Fallon's parents moved to the U.S. from the Dominican Republic when Alma was nine. After high school, she worked as a bank teller—just like her older sister—for several years before going to college to study engineering. She's standing here next to a model of the CVN-21 aircraft carrier.

Gateway to the East

Look around you and a huge portion of the consumer goods you see—like furniture, clothing, electronics, toys, and computers—were made in Asia. How do they get to the U.S. from China or Japan or Korea? Mostly through the Port of Los Angeles.

Over $120 billion worth of consumer goods move through the Port each year! The Port of L.A. is the busiest port in the U.S., and eighth busiest in the world. The Port's harbor, piers, cranes, and container terminals are world class—and civil engineer **STACEY JONES** (b.1956) is at the helm.

"We were successful in building Pier 400 because we were a team," says Stacey Jones, shown here under the towering cranes that move 2.4 million TEUs (twenty-foot-equivalent units) of cargo every year.

Stacey is the Port of Los Angeles's director of engineering development. One of Stacey's most important projects was Pier 400—in which sediment was dredged from the harbor and used to create new land. With Pier 400 built, the Port was able to attract Maersk-Sealand, one of the largest shipping companies in the world, to call the Port of L.A. "home."

Endangered California least terns also call Pier 400 "home." The Port of L.A. maintains a protected nesting site at Pier 400 by grading, removing vegetation, placing decoys, and providing chick shelters. The number of nesting pairs and fledglings has been increasing every year.

Join the Web of Women Engineers

When asked why women make good transportation engineers, **AFSANEH "SUNNIE" HOUSE** (b.1958) replied, "Women make good anythings, not just engineers!"

Sunnie points out that although engineering is all about approaching a problem in a logical way, it's also about judgment. "Good judgment comes from instinct—and women have great instinct!"

Sunnie is a transportation engineer and is the 2004–2006 president of WTS, the Women's Transportation Seminar, an international association of women and men in all fields of transportation. One of WTS's primary goals is to help advance women in the industry.

"The biggest value that WTS, SWE (the Society of Women Engineers), and other groups provide to women is a really strong network. We have resources and we're here to help women succeed in their careers."

Growing up in Iran, Afsaneh "Sunnie" House knew that she had to follow her father's wishes when it came to choosing her life's path. At the time it was quite unusual for a traditional Iranian father to encourage his daughter to get an engineering degree. But he did, and she did, and Sunnie has enjoyed a successful 21-year career as a transportation engineer.

That's Entertainment

Entertainment makes life magical and downright fun! Entertainment gives life a lift! Who's behind almost all of today's modern entertainment? Women engineers, of course!

What kind of engineers are they? Well, mechanical engineers create toys, sports equipment, gadgets, and gizmos—parts to make things work. Civil engineers create stadiums, resorts, museums, zoos, concert halls, and theatres . . . not to mention shopping malls and libraries.

Electrical engineers create the technology behind streaming live concerts, television, radio, digital

Lightning Racer, Hersheypark, Pennsylvania.

cameras, and virtual reality. Computer engineers create the fantastic animation and special effects used in movies, videos, artificial intelligence, and computer games . . . along with circuses and Broadway shows.

Together, they work as "wizards of wonder" to deliver magical marvels that stir our imaginations and provide us with laughter, leisure, and hours of fun.

> *"Too much of a good thing can be wonderful."*
>
> MAE WEST
> ACTOR (1893–1980)

Destination: Fun

What's your favorite destination? Grandmother's house? The beach? A theme park or a ballpark? How about an aquarium, vivarium, or zoo?

Leisure travel accounts for over 80 percent of the person-trips made in the U.S.! Shopping is the number one activity, followed by social or family events, and then outdoor activities.

Traveling lets us see new sights, experience new things, be with new people, and kick back and have fun! Safe and marvelous fun, thanks to women engineers.

Erin Wallace helped create and develop Disney's Typhoon Lagoon Water Park at the Walt Disney World Resort in Orlando, Florida. The breakthrough water park, the first with a theme, opened in 1989 and has been an enormous success ever since. Erin recalls the enjoyment of traveling to various water parks to study layouts, rides, and guest patterns before analyzing all the information to build Disney's Typhoon Lagoon.

The Magical World of Disney

Walt Disney World® opened in Orlando, Florida, in October 1971, and continues to delight millions of visitors each year. It contains 47 square miles of world-class entertainment, including four theme parks, two water parks, 32 resort hotels, golf courses, spas, a sports complex, Downtown Disney—an entertainment-shopping-dining complex—and more. It's big enough to be its own city!

Industrial engineer **ERIN J. WALLACE** (b.1959) has an equally big job. She's responsible for making all of Walt Disney World work smoothly.

"Guest-ology," the study of how guests use the park, is her major focus. Erin examines and measures how guests use attractions, food, and transportation to determine the best operating procedures, and to develop the best proportion of rides and recreation to meet demand.

Disney has extremely high standards for the treatment guests receive—standards that make Walt Disney World the dream world millions love to visit. "Industrial engineers have become such integral and valued partners in the creation of Walt Disney World that lots of them are hired," Erin notes.

There's an endless array of intriguing problems to solve, and Erin plays a very major role in finding marvelous—and magical—solutions.

The Scream Machine

The Mauch Chunk Switchback Railway in Pennsylvania is considered the first roller coaster in the U.S. The company wanted revenue from their idle mining cars, so it invited people to clamor aboard for the nine-mile, downhill ride.

Today, roller coasters spiral, twist, roll, corkscrew, loop, and pretzel up to 60 miles per hour! Mechanical engineer **ROBBIN S. J. FINNERTY** (b.1959) knows all about roller coasters: she builds them! Robbin makes sure the wooden coasters built by Great Coasters International are stable and safe.

A coaster in action can exert thousands of pounds of force on the structure, and on the connections to the foundation. As Robbin works with the coaster designer to build the supporting structure, she has to be aware of how much force these critical connections can handle.

Wood can be a tricky building material for coasters since it changes shape depending on temperature and humidity. It's up to Robbin and crew to produce a "tight woodie."

There are basically two kinds of roller coasters: wood and steel. Wood coasters don't usually loop, aren't as high, and go slower than their steel cousins. "But they have a feel and sound many people find irresistible. After the initial ride up the lift, the laws of physics take over, and gravity sends the coaster speeding downwards," says Robbin Finnerty. "The higher the drop, the more energy there is to play with. The more energy, the longer or faster the ride."

Above: "Thunderhead, built in the foothills of the Tennessee Smokey Mountains, was the biggest challenge so far," says Robbin Finnerty. The coaster has a 100-foot drop and speeds of 54 miles per hour! A segment of the track is shown here. Typically, coasters are built on fairly level terrain or on concrete slabs, often with steel supports. Thunderhead was built in a steep valley. Robbin collaborated with other engineers and ended up driving steel pilings into the hillside, and building a stepped concrete supporting foundation.

At left: PowerPark, a popular resort destination in Harma, Finland, is home to a new wooden coaster, shown in this rendering, which depicts the intricate structural skeleton and track twists (in red).

TAKE ME OUT TO THE BALL GAME

As fans know, baseball games are canceled if it rains. The old Seattle Kingdome was covered to keep out the rain, but there's nothing better than watching baseball outdoors on a sunny day.

So when the new stadium, Safeco Field, was built to replace the Kingdome, it was designed with a retractable roof and sides that can slide open or closed, depending on the weather. Now Safeco Field has the best of both worlds!

During construction of the roof, a steel platform was positioned up to 250 feet above the ground for the crew to stand on while they worked. Civil engineer **CHERYL M. BURWELL** (b.1975) developed a computer model of the platform and analyzed the impact of wind gusts to determine how far the platform could sway before it became too dangerous for the construction workers.

SAYS CHERYL BURWELL, "I GET TO WORK ON DESIGNING BUILDINGS ON PAPER, THEN SEE THEM BECOME REAL FOR PEOPLE TO ENJOY. ALL THE PROJECTS ARE SO EXCITING. IT'S TOUGH NOT HAVING ENOUGH TIME TO WORK ON EVERY ONE."

AT RIGHT: BASEBALL FANS ENJOY AN EVENING IN THE OPEN AIR AT SAFECO FIELD WHERE THE RETRACTABLE ROOF AND SIDES OF THE STADIUM HAVE BEEN ROLLED BACK.

It All Falls Down . . .

CHERYL BURWELL also played an exciting role during the implosion of Seattle's old Kingdome Stadium in 2000. Before the implosion, all the buildings surrounding the stadium had to be evacuated so no one would get hurt. Cheryl and her team monitored buildings for movement during the blast. Immediately following the blast, while dust and debris were still falling, she was out checking accelerometers that measure the earth shaking, before her colleagues could examine buildings to let people know when it was safe to return. "With all the dust generated from the implosion, the landscape was surreal, but the job was so cool," Cheryl recalls.

AQUATIC HOMES

"Animals in zoos and aquariums need a healthy home, too! That's especially true for marine animals in aquariums since they are very sensitive to water quality," says civil and environmental engineer **MILICA KALUDJERSKI** (b.1970).

Milica is an aquarium expert. She designs water, recycled water, and wastewater treatment systems, as well as life support systems, for fish and aquatic animals. "There have been many advances in water treatment, but there's still a lot to learn about protecting this important resource for both people and animals."

When the sharks, turtles, penguins, and live coral had to move from their home at the California Academy of Sciences in San Francisco during a building remodel, Milica created 15 tanks that held from 1,000 up to 21,000 gallons of water.

"Water in an aquarium must contain nutrients to feed the fish. But waste build-up must be prevented, so water is constantly recirculated through filtration and disinfection systems. Also, water must remain clear to keep animals healthy and allow the public to see them."

MILICA KALUDJERSKI EXAMINES THE WATER QUALITY AND CLARITY IN ONE OF THE AQUARIUMS FOR WHICH SHE DESIGNED THE FILTERING AND TREATMENT SYSTEMS.

LILY WANG IS ALSO STUDYING TIGERS, MEASURING THE SOUND FREQUENCIES OF THEIR ROARS, CHUFFING (LIKE PURRING), AND MATING CALLS, TO BETTER UNDERSTAND THEIR COMMUNICATION PATTERNS. LILY HOPES TO USE THIS INFORMATION TO HELP TIGERS SURVIVE AS THEIR TERRITORY IS ENCROACHED UPON BY HUMANS.

GOOD VIBRATIONS

Ever wonder why music sounds so great in a concert hall? It's because of acoustics: how sound is generated, transmitted, and received. Room size and shape, construction materials used, and their placement all affect sound reverberations and quality.

Architectural engineer **LILY M. WANG** (b.1971) studies acoustics in concert hall design. She's been working on acoustic predictions for the new Holland Performing Arts Center in Omaha, Nebraska. "I've used a scaled architectural model of the building and a 3-D computer model to test the distribution and reverberation of sound throughout the hall," Lily explains.

When the new symphony hall is completed, she'll measure the actual acoustics in the building to compare against her two models. "I'm helping to improve the acoustic modeling techniques so that future concert hall designers can accurately plan for the sound quality they want while the building designs are still on paper."

Lily used to dream about designing concert halls—so her job is truly a dream come true! "I go to concerts and listen carefully to the sound. That's so cool."

HARDWIRED TO PLAY

Playing isn't just for kids. Research has shown that play is a fundamental need in the lives of humans and other species. It keeps us physically and mentally fit. It helps us learn and promotes creativity. Play fosters belonging and cooperation. So bring on the toys and games!

Dolls, models, toys, scooters, skateboards: they're all engineered! Design engineers help develop the concept, molding engineers build form-fitting molds for parts, product integrity engineers test toys for safety and reliability, and packaging engineers figure out how to fold and fit everything into the box.

When it comes to video games, engineers have helped create an $18-billion-a-year gaming industry that surpasses movie box office ticket sales and keeps on growing.

Engineering advancements in entertainment are also paving the way for artificial intelligence and its use in toys, movies, and household applications.

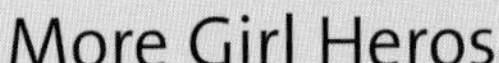

More Girl Heros

Video games may be girls' first introduction to computers, and can help them become comfortable with technology they'll use in school. But, almost 70 percent of video gamers are male. Why? Research shows girls generally don't find violence or cutthroat competition appealing.

New games are needed for girls:

- Collaborative games with characters of interest to both genders.
- Games that have more girl heroes.
- Games that use intelligence and compassion to surmount challenges.

REAL OR ROBOT?

"Imagine having a fanciful pet that's really a robot!" exclaims electrical engineer **CYNTHIA BREAZEAL** (b.1967).

Sociable robots aren't science fiction any more. Cynthia consulted on the movie *A.I.* (Artificial Intelligence) and sees sociable robots having domestic and healthcare applications as well.

She's currently working on "Leonardo," a completely autonomous sociable robot with "65 motors or degrees of freedom." These enable some speech capabilities, gestures—like blinking or smiling—and other movements. "We don't realize it, but we have very complex interactive social skills that can't be easily programmed into robots."

"Teddy," the robot teddy bear in *A.I.*, is actually a robot puppet. It needed five human puppeteers to control its movements. Says Cynthia, "It's hard to coordinate a team of people to make a robotic puppet gesture while maintaining eye contact. But an autonomous robot can do this on its own." She adds, "You could say I'm making real robots into better actors."

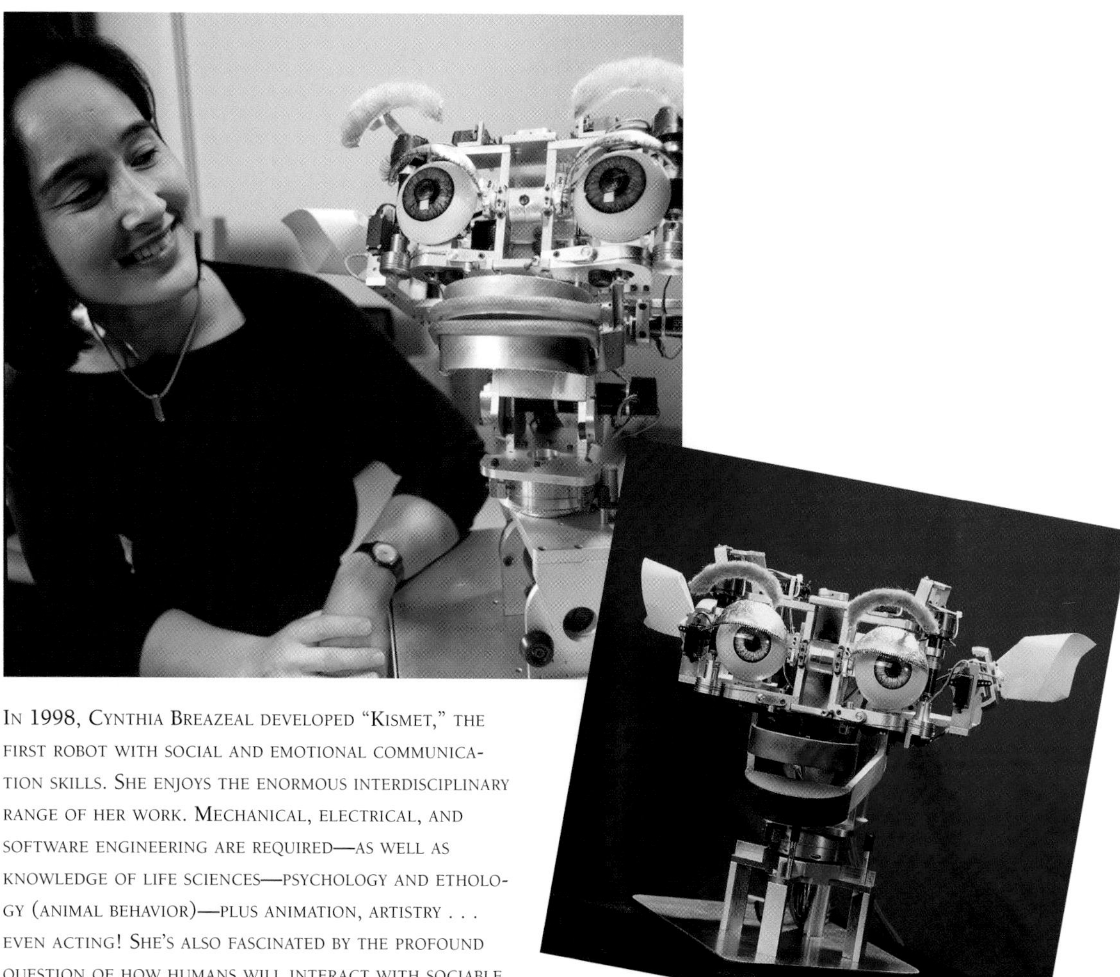

IN 1998, CYNTHIA BREAZEAL DEVELOPED "KISMET," THE FIRST ROBOT WITH SOCIAL AND EMOTIONAL COMMUNICATION SKILLS. SHE ENJOYS THE ENORMOUS INTERDISCIPLINARY RANGE OF HER WORK. MECHANICAL, ELECTRICAL, AND SOFTWARE ENGINEERING ARE REQUIRED—AS WELL AS KNOWLEDGE OF LIFE SCIENCES—PSYCHOLOGY AND ETHOLOGY (ANIMAL BEHAVIOR)—PLUS ANIMATION, ARTISTRY . . . EVEN ACTING! SHE'S ALSO FASCINATED BY THE PROFOUND QUESTION OF HOW HUMANS WILL INTERACT WITH SOCIABLE ROBOTS.

BELOW: KISMET MUGS FOR THE CAMERA.

DOLLS FOR TWEENS

Toy companies call girls ages 8 to 14 "Tweens," and creating toys for Tweens is big business. In 2001, toy-maker Mattel released the Diva Starz,™ four stylish, sassy, high-energy teenaged dolls, each with her own personality, intended for the Tween market.

The interactive Diva Starz contain clocks, contacts on their bodies, and electronic tags in their clothing and accessories. The doll's computer reads the tags in her clothes so it knows what she's wearing. All this gadgetry lets the Diva Starz talk about their wardrobes and activities at different times of day. Infrared technology in their feet enables two dolls facing each other to "converse."

A PLANT IN CHINA MANUFACTURES DIVA STARZ DOLLS. ALLISON CONNER, STANDING IN THE MIDDLE OF THE ASSEMBLY PLANT FLOOR, SAYS, "INDUSTRIAL DESIGNERS CAN COME UP WITH IDEAS THAT AREN'T ALWAYS REALISTIC. MY JOB IS TO MAKE THE PRODUCT SAFE, MANUFACTURABLE, AND COST-EFFECTIVE WHILE REMAINING AS TRUE TO THE ORIGINAL CONCEPT AS POSSIBLE."

Mechanical engineer **ALLISON CONNER** (b.1976) worked on Diva Starz parts such as snap-on clothing, and accessories such as cell phones. "Working with industrial designers to make toys for little girls was a lot of fun," Allison says.

Allison's job begins after she receives a design. Using 3-D design software, Allison takes a product design, breaks it into parts, and works with tooling vendors to develop and refine the pieces. "If you enjoy making things and solving problems in creative ways, this is a great job," she says.

TEENY DOLLS

In the last 15 years, Mattel's Polly Pocket™ doll has evolved from a one-inch-tall doll to a three-and-a-half-inch-tall plastic fashion doll with clothes, vehicles, and real estate anyone would envy. Mechanical engineer **CONNIE DALE** (b.1974) has been one step ahead of market demand, designing fabulous new Polly Pocket accessories girls love to play with.

"People think toy engineering should be simple, but it's actually very challenging. There are so many time, cost, and material constraints on what can be produced. Tremendous creativity and efficiency are required, but it's fun," says Connie, who loves the fast-paced environment. She's constantly working with a team of people and interacting with many different departments to create the toys.

Connie started working in the toy business in college. "My college had a mandatory internship program, so I happened to work with toy-maker Hasbro in summer and winter. I did that for three years and it helped me land a job with Mattel after I graduated. I've been working on toys ever since."

"DESIGNING TOYS KEEPS ME YOUNG," SAYS CONNIE DALE. "I SPEND A LOT OF TIME INTERACTING WITH LITTLE GIRLS, SO I LEARN A LOT ABOUT HOW THEY'RE GROWING UP IN THE WORLD."

Create and Conquer

Creating and managing the development of video games is the work electrical engineer **ATUSSA SIMON** (b.1979) loves to do. "I work a lot with companies developing their brands, so the work I do is largely confidential," says Atussa.

Atussa uses her skills to keep software engineers, artists, content developers, environment engineers (who develop the game setting, clues, and switches), and voice actors on track and within budget to deliver the final game by the deadline. "Bringing the team together harmoniously to create a game is a very rewarding experience."

Atussa is currently using machinima (machine animation) to create computer game graphics that are almost as refined as computer graphics (CG) movies. She finds it especially challenging to develop and draw emotion out of non-human characters in a convincing way.

"Playing the game is like being part of the movie. In fact, machinima is also a great way for movie-makers to test a scene inexpensively without having to build sets and test camera angles." Atussa foresees video games and movies coming closer together in the future.

Atussa Simon wrote her first computer game at age six and started a game company in seventh grade to sell puzzles. "I developed my first serious video game called *Mythica,* with a group of girls . . . just to prove that women could produce and design a serious game," comments Atussa, who notes the industry is largely dominated by men, though that's starting to change.

"To love what you do

and feel that it matters—

how could anything be more fun?"

Katharine Graham

Publisher of The Washington Post (1917–2001)

The Moving Picture Show

Who doesn't like watching movies, videos, or DVDs? Engineers have contributed huge gains to technology that affects animation and special effects, not to mention cameras and other equipment, light systems, audio systems, color, set building, and much more.

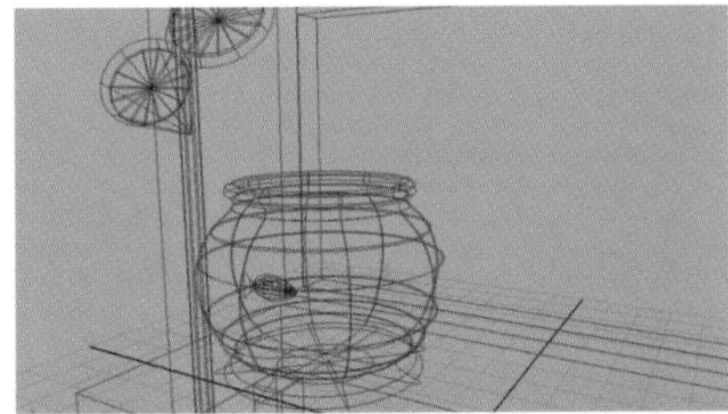

To create computer graphics, a wire frame is first developed using complex mathematical formulae to determine the surface of the objects. Next, basic colors are applied to the same scene. Finally, computers use shading programs to render a realistic-looking image with texture and depth.

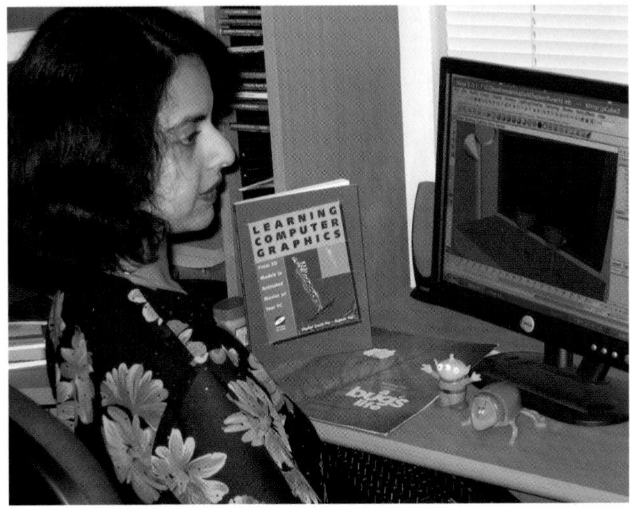

Light-Year Graphics

In 1995, the public flocked to movie theatres to enjoy the first full-length 3-D animation film, *Toy Story,* which not only charmed children and adults alike, but took the animation world by storm.

Computer engineer **SHALINI GOVIL-PAI** (b.1969) was technical director on *Toy Story,* developing tools like Marionette that enabled artists to animate characters on a computer, without being aware of all the "deep math" algorithms cranking in the background. Marionette later won its own (scientific) Academy Award®!

"I used to think technology was all-powerful in itself. I love the sophisticated math behind it. But when I worked with creative artists and animators, I realized that all the brilliant math and technology was useless if they couldn't use it. Simplicity is key," says Shalini.

Shalini was also technical director on *A Bug's Life* (1998), and now runs her own company making computer graphics characters for Internet movies and games. "When I was 13, my father got one of the first computers for graphics in India. I loved it. I knew then computer graphics was the career I wanted to pursue."

Shalini Govil-Pai was technical director on the blockbuster films *Toy Story* and *A Bug's Life*. She has developed several virtual reality-based games and flight simulators. She's also directed commercials for major clients such as Nabisco, Levi Strauss & Co., and McDonald's, and has authored several books on computer graphics. "Engineering is the most thought-provoking career there is."

The New Cast of Thousands

Biomechanical engineer **SIÂN E. M. JENKINS LAWSON** (b.1977) loved horses as a child, and loves them still. She now animates horses, animals, and humans for big-budget movies like *Troy, King Arthur, Alexander 2004,* and *Kingdom of Heaven.*

As a teenager, Siân seriously considered a career in show jumping, but reluctantly went off to college where she studied math before finding a program in Equine Science.

While studying biomechanics at Oxford University, Siân used "motion capture" to model children with walking difficulties. "The motion capture camera I used to track movement in children was the same kind used in movies, so I told kids to pretend Spielberg was filming them."

Motion capture is a way to animate the movement of digital characters—animals or human beings. Reflective markers are placed on crucial skeletal and muscle points. A camera records the markers while the subject moves. 3-D coordinates are then created from the recording and input into a computer model that portrays the data as a moving skeleton.

Animators later add skin, hair, and clothing to make realistic beings, or they can wrap a photo image of the animal or person around the 3-D image on the computer to produce action-packed movie scenes without the cost of filming.

Using motion capture for the movie *Troy,* Siân created roughly 100,000 Greeks and 40,000 Trojans with horses, weapons, and ships in about two months!

Why does Siân do this? She's a strong advocate for the humane treatment of animals in the entertainment industry. Siân has developed simulation models and tools to animate horses and other animals so they are not subjected to the traumas of movie-making.

"Many directors still want horses to race towards the camera and fall over," Siân says. "To get the effect, they install trip wires, which cause injury to the horses. Using realistic simulation will eliminate this issue altogether."

Siân is now taking horse animation to the next level: she's using motion capture to study how a horse's skeleton, muscles, and tendons move under the stress caused by different kinds of horse-shoes.

Ever persistent, Siân knocked on many doors until she found her niche in movie-making. Siân now owns her own company, Equine Mechanics, with her husband. "We go to movie locations together and bring our child."

"I admit I've found my dream job. It's fun and interesting, and I'm helping horses and other animals."

—Siân Lawson

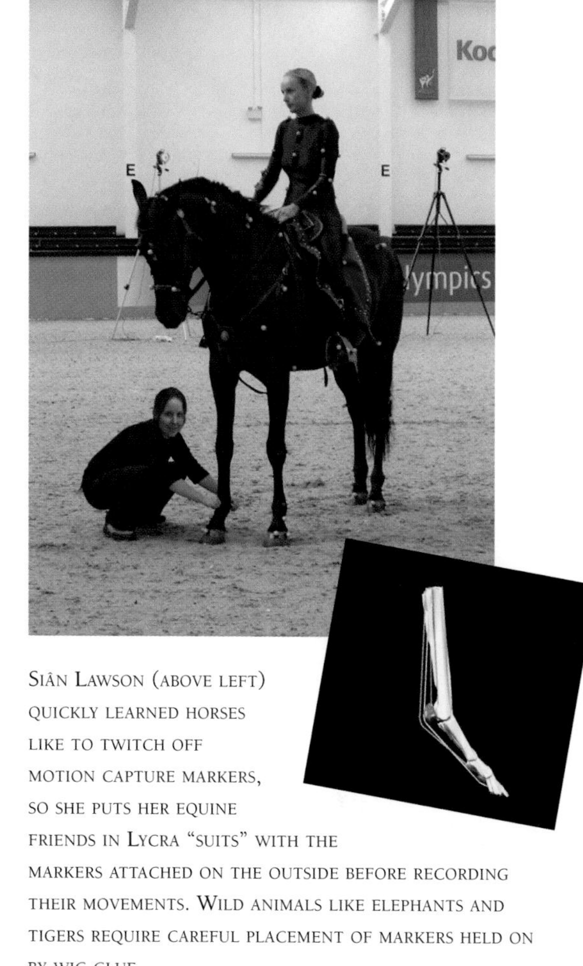

Siân Lawson (above left) quickly learned horses like to twitch off motion capture markers, so she puts her equine friends in Lycra "suits" with the markers attached on the outside before recording their movements. Wild animals like elephants and tigers require careful placement of markers held on by wig glue.

Bottom: This image shows a computer-generated skeleton of a horse's leg as well as the stretch of muscles and tendons. Motion capture markers (in red) are positioned on the live horse to provide coordinates for the camera to record while the horse moves. 3-D coordinates from the recording are then used to create a moving skeleton in the computer.

Hollywood Tech Star: A Day in the Life of Wendy Aylsworth

Hollywood produces special effects galore—mostly on computers. But the movies we see are still largely shot the old-fashioned way on celluloid film. It's time movies went digital!

Computer engineer and technical guru Wendy Lynn Aylsworth is leading the charge. As vice president of technology at Warner Bros. Studios, she oversees the hottest new technology for TV, movies, and Digital Cinema (D-Cinema).

Over time, celluloid film becomes dusty, scratched, and faded, but D-Cinema keeps movies looking as perfect as they did on opening night, even after hundreds of viewings! Some day soon, you'll see digital movies with more vibrant colors and higher-resolution images on screen for a richer movie-going experience.

Wendy Aylsworth is standing in front of a set from *ER* on the Warner Bros. lot where she works on new technology used in TV and movie production. Wendy's helping to standardize production technology—getting all the bits and parts to communicate between each other in a methodical way—so TV programs like *ER* can be seen in many countries.

How does D-Cinema work? It begins with a digital film master that's created using a digital camera, computer-based animation, or film that's been digitized on a film scanner. Digital audio is then added before the digitized film is compressed and sent to the movie theater via disks or satellite. Once it's received at the theater, digitized content is loaded into a digital projector for viewing. Here, Wendy is checking the quality of a digital master.

When she's not traveling on business, Wendy spends much of her day in meetings figuring out how to put new technology into place behind the scenes. Here, Wendy strolls on the set of the *Gilmore Girls* en route to a planning meeting for a D-Cinema launch, though "we never interrupt during a shoot."

"Troubleshooting technical problems is part of every workday. As technology becomes more complex, problem solving requires different creative approaches." For example, Wendy's team is testing one of the first 4K digital projectors (below) in the world. "We ran a movie on it and suddenly the screen went black for a moment. We had to figure out why the system hiccuped." Warner Bros. has begun a one-year trial of a digital movie system in Japan. The first film to run is Tim Burton's *Corpse Bride*.

Wendy frequently travels abroad to attend conferences on standards for TV and movie technology. In fact, she sits on a United Nations committee for international telecommunications. "When we standardize technology, we can share entertainment all over the world. For example, North American TV programs and movies can be shown in Asia, and vice versa. These trips are exciting and give me a chance to meet interesting people, see other cultures—and do some shopping."

When Wendy's at home, she enjoys watching a softball game with friends and family. Sometimes she even plays on a team. On weekends, she often catches up with her college-age sons. "I recently spent a Saturday painting dorm rooms and moving furniture!"

What's Wendy's other passion? "Making a good education accessible to all kids! A good education teaches you how to get information to make the best decisions. That's a critical skill to have throughout life."

Entertaining Connections

"Broadband for all" some entertainment visionaries are predicting. Broadband, or high-speed digital data delivery, may make it possible to combine television, computer data, Internet, and phone service into one package.

Convergence of all these services, for those who want it, is not too far away. That means even more entertainment at our fingertips!

Coming to You Live!

When electrical engineer **WANDA GASS** (b.1956) joined Texas Instruments in 1980, she began working on what was then new technology—digital signal processing (DSP). DSPs receive audio or video signals that have been converted to digital format, manipulate them, and then transmit the signals so fast that the sound or pictures seem "live."

Wanda recalls that one of the first applications for DSP technology was a doll that could make conversation. "Certain key words were programmed into the doll's internal computer. She would prompt me with questions using keywords, then process my response using DSP to receive my voice data signals. By using DSP to transmit back bits of conversation based on the key words, we could have a 'live' conversation."

Now DSP image and audio applications are used extensively for cell phones, digital cameras, MP3 players, and PDAs (personal digital assistants), to name a few. While a microprocessor can do number crunching, it's not fast enough to stream video, movies, or music. That's where DSP is critical. "We're also looking at future trends, like using DSP technology in small chips to wear in your clothes so you can have access to wireless communication anywhere."

Right: A Digital Signal Processor (DSP) is a very fast and powerful specialized microprocessor that processes data in real time.

Far right: "Engineering allows you to follow your dream, invent your passion, and channel that passion into something practical," says Wanda Gass. "You can affect many people's lives. My ninth-grade daughter asks why I'm concerned. My answer is that we can make better products if women are engineers, too. Diversity is important. Everybody adds something different to a project, and that's what makes the best product."

Background: XM satellite dish.

The Next Cool Thing

"I thought I wanted to go into chemical engineering," says **SHERITA T. CEASAR** (b.1959), "but the smells from the lab caused me to change engineering discipline." After scoring high in mechanical skills on an aptitude test, she chose mechanical engineering instead.

Now Sherita is a leader in the field of broadband communication services for digital cable TV, high-definition TV (HDTV), high-speed Internet access, video-on-demand, and telephone service.

"As a kid I was always fixing things. Now I manage engineers who make all this possible." Sherita's referring to the creative minds that determine how TVs can digitally record programs, capture and store programming for playback on demand, fast-forward through commercials, surf the Internet at high speeds, and always be reliable.

"Our engineers are constantly searching for the next thing. I want to develop a new level of customer service, so people can rely on continuous performance. For that, I still use the systematic problem-solving approach I learned in engineering."

"If something doesn't work, make another choice," says Sherita Ceasar. "It's important to make choices that will give you more interesting choices down the line." In high school, Sherita participated in a minority program to improve her math and science skills, then interned at General Electric during summers. "I got to sample lots of engineering disciplines and pick the best one for me."

"When a satellite signal is blocked by tall buildings or mountains, we use a repeater to amplify and re-broadcast the signal," says Michelle Tortolani. Here, she's checking a repeater to make sure it's processing signals effectively.

Tune Into XM

The number of stations on "terrestrial" radio, one of the few remaining analog communications services in our digital world, is limited. Satellite radio—and TV—can provide hundreds of channels, which offer a huge range of programming choices.

The XM system provides digital radio broadcast service via satellites and terrestrial repeaters to consumers in the U.S. People listen through specially designed receivers in cars, trucks, and recreation vehicles, as well as through portable and home-based receivers. In areas where XM satellite reception is obstructed, a network of ground-based repeaters broadcasts the programs.

Electrical engineer **MICHELLE TORTOLANI** (b.1960) managed the design, development, and manufacturing of the repeater hardware for the XM system. "With satellite radio, when you can see the sky, you're getting the signal. But in a downtown area, buildings can block your signal. So XM had to come up with a way to receive, amplify, and re-broadcast the satellite signal."

Says Michelle, who currently manages the operation and maintenance of the nationwide repeater network in more than 60 cities, "XM has given me the opportunity to be on the forefront of the development and launch of an exciting new technology—an engineer's dream!"

Let's Communicate

Ask a woman in 1900, "How do you stay connected with family and friends?" She'd say letter-writing, but sometimes it took weeks for her letters to be delivered. Telegrams were only for major emergencies.

Radio, or "wireless," as they called it back then, had just been invented, so it would be 20 years before she and her family gathered 'round the radio to listen to music, news, comedy, and baseball.

TV IN THE 1950s.

Ask the same question of a woman in 1950, and she'd point to her rotary telephone. And all her friends were buying televisions. She wanted one!

Today? A woman has all sorts of inexpensive options: land-lines, cell-phones, e-mail, and the Internet, along with television, radio and, yes, we still have telegrams. Never before has a woman had so many ways to communicate easily, clearly, and quickly!

Engineers are now putting it all together, a convergence of capabilities into one "anytime, anywhere" window to talk "whenever, wherever" we want.

We Need to Talk!

In the early days of telecommunications, women primarily served as technologists—meaning that they assembled, worked on, and maintained technology rather than engineering it.

Mattie "Ma" Kiley (1880–1970), for example, became a famous "telegrapher" who mastered Morse code and telegraph operation for the railroads. Women like Ma changed the general perception that women had no mechanical or electronic capabilities.

By the 1960s, our telecommunications infrastructure had both speed and range. Intercity traffic was carried by radio frequency between towers, and by copper wire cables. Intercontinental traffic went by undersea copper cables. Not long after, satellites helped transmit telephone and high-speed data such as television broadcasts from the sky.

Today, the vast bulk of communications around the world rely on fiber optic cables that can carry far more traffic, making worldwide communication affordable to everyone.

Background above: Morse code translation of the words on this page.

Right: Long Distance telephone exchange in London.

Thousands of Calls

In the 1930s and 1940s, when you picked up the phone, an operator said, "Number please?" The operator then plugged wires into a switchboard and your call went through. This was a great system for a few hundred calls a day—only if you had a pretty good staff of operators. But imagine if you had thousands of calls!

By 1950, most telephone calls were processed by telephone switches that were able to detect when a caller lifted the telephone off the hook, counted the numbers dialed, figured out where the wires of the two parties appeared on the network, connected the call, and later disconnected it.

Bell Telephone Laboratories knew it needed to improve the system to support the anticipated demand. **ERNA SCHNEIDER HOOVER** (b.1926), a mathematician and computer programmer, developed a computerized system that monitored incoming calls and then automatically adjusted the call's acceptance rate. This helped eliminate overloading problems.

Communication companies today still rely on the fundamental principles of Erna's switching system. It's even the precursor for modern e-mail routing systems!

In her spare time, Erna Hoover enjoys spending time in the outdoors, skiing in the winter and hiking in the summer with her husband Charles.

Erna Hoover drew up the first sketches of her telephone switching system while in the hospital after giving birth to one of her three daughters. Her switch was able to automatically handle thousands of calls an hour and held up under unusually heavy calling when some exciting event occurred, such as a natural disaster.

Even though she didn't have a formal engineering education, Catherine Eiden's experience earned her the position of Staff Engineer at Illinois Bell. Her estimating methods became a *de facto* standard throughout the company.

Millions of Phones

As telephone companies grew in their ability to handle a greater number of customers, so did the demand for service beyond the major cities. The phone companies needed more poles, wires, transmitters, and other basic equipment.

A new engineering expertise had begun to emerge: an engineering estimator—a person who understood the telephone technology, the needs of the customer, and how to best balance the two for good service and profitability. **CATHERINE EIDEN** (1914–1990) was one of the best.

Catherine went to work in the Illinois Bell Telephone Company's engineering department as a clerk when she was 20 years old. With a quick mind and good math skills, she taught herself the technical side of telephony and combined that with her knowledge of business and people. Catherine analyzed existing systems, compared their capabilities against projected customer demand, and then recommended the best way to expand the system.

AT A YOUNG AGE, EDITH MARTIN HAD THIS INTERPRETATION OF THE U.S. CONSTITUTION: "IN THE UNITED STATES OF AMERICA, YOU CAN DO OR BE OR HAVE WHATEVER YOU ARE WILLING TO EXERT THE ENERGY TO ACCOMPLISH." SHE USED THIS PHILOSOPHICAL VALUE TO GUIDE HER THROUGHOUT HER LIFE AS A TECHNOLOGIST, A LEADER, AND AS A MOTHER OF TWO. SHE GIVES THIS ADVICE TO PEOPLE: "KNOW YOURSELF. KNOW WHAT'S IMPORTANT. AND THEN EVERY YEAR DO SOMETHING THAT MAKES YOU MORE WORTH KNOWING."

FROM SAVVY SATELLITES TO DIGITAL DOMINANCE

Beginning in the 1960s, as computer and electronics technology developed, engineers began transforming the telephone into an invaluable multi-functional tool, thanks to the increasing number of communication satellites. Like a global net, these satellites made it easier for people all over the world to stay connected.

For many years, computer scientist **EDITH W. MARTIN** (b.1945) made sure that the satellites operated reliably and that the Earth-based monitoring stations received and disseminated information as efficiently as possible. She helped the company establish high-speed Internet capabilities for customers around the world, increase the number of satellites for more comprehensive service, and make webcasts and tele-conferences an affordable part of business life.

Prior to this, Edith helped aircraft manufacturer and defense contractor, Boeing, keeps its technology edge in commercial and military aviation. In particular, she built, staffed, and managed Boeing's $150 million High Technology Center, an innovative research facility that prototyped some of the world's first technology devices that included solar cells for the Space Shuttle and microwave integrated circuits that create secure communication systems. Today, Edith is one of the leaders of a company that wants to deliver a new age of communications that she says will change the way people interact with the digital world.

Edith says, "I've always wanted to make a difference. To use my technical expertise and management abilities to facilitate change. My philosophy is if you can envision it, then you can do it."

Signals From Above

On a clear moonless evening, take a look at the night sky. If you watch carefully, you'll see something that looks like a star, except it's moving.

Chances are this is a satellite, traveling across the horizon several hundred miles away. While you're watching, it's sending information back to Earth.

There are an estimated 3,000 satellites owned by 40 different countries currently orbiting the Earth. Many of these satellites are there to provide entertainment, news, and communications to people all over the world.

One of them may be "beaming down" your favorite television show or movie, or helping you talk on the phone to friends or family.

Many others are used to predict weather, study the environment, navigate to within a millimeter of where you're standing, manage military operations, or even search deep space.

SCIENTISTS LAUNCHING A SATELLITE FROM THE SPACE SHUTTLE.

With design and development complete, Yvonne Brill inspects her hydrazine resistojet propulsion system just before its launch into orbit. When the patented system was revealed and people realized what a simple and practical design Yvonne had come up with, one colleague remarked, "If it was so simple, why didn't a man invent it." Yvonne just smiled and went back to work on her next propulsion system.

Engines to Orbit

Once launched into the Earth's orbit, satellites rely on small rocket engines to keep them in place. In the fledgling days of satellite development, propulsion engineer **YVONNE BRILL** (b.1924) designed a new rocket engine, called the hydrazine resistojet propulsion system. Because it uses 35 percent less fuel than other engines, it enables satellites to stay in orbit longer or carry more payload.

Yvonne's engine is now standard on all RCA (now called GE Astro) communication satellites. Yvonne says, "I was just fascinated with rockets and missile designs, and particularly the propellants that allow these rockets to fly into space. I always looked to push the envelope and come up with new ideas that would get the spacecraft, satellites, or whatever, safely and reliably into space."

Yvonne says that while her formal education is in math and chemistry, it's her on-the-job training that turned her into an engineer. "Engineering holds so many possibilities. It's the chance to take theory and put it to work in real-world applications. That's what we did: take some basic knowledge about rocket fuels and engines, and figure out how to put a satellite in orbit—and keep it there."

THE INFORMATION AGE

Early computers were electronic brains, built to solve problems quickly, not to communicate. However, as computers advanced from vacuum tubes to integrated circuits to microprocessors (inside all computers today), the way people could use computers also progressed.

A computer is still a huge electronic brain—though "smarter" than ever, especially now that computers are synonymous with the Internet, e-mail, and instant messaging. Just 10 years ago, these tools didn't even exist!

"Humans are allergic to change. They love to say, 'We've always done it this way.' I try to fight that. That's why I have a clock on my wall that runs counter-clockwise."

—GRACE HOPPER

SAY IT IN ENGLISH!

In the 1950s, computer scientist **GRACE MURRAY HOPPER** (1906–1992) commented that a lot of people didn't understand computer language and a lot of computers didn't understand human language.

In fact, every piece of software was written in binary code (computer language in 0's and 1's). Creating software was tedious, time-consuming, and more than a little frustrating.

A true visionary, Grace believed that the viability of computers depended on the development of languages that could be understood and used by people who were not computer experts.

Grace suggested that UNIVAC, the first large-scale electronic digital computer for business, be programmed to recognize English commands. Despite ridicule from her peers, Grace spent many months building a compiler —the link between the person writing the code (or the program) and the hardware of the machine network.

Grace's compiler could understand and translate specific English statements to perform common business tasks such as payroll calculation and automated billing. "Everybody said we could not do that. So we went ahead and did it," she recalls.

She then went on to develop the Common Business Oriented Language (COBOL), a computer language that could be used to create software. COBOL was a remarkable leap forward in the computing industry. Programmers now had a common language to build business, financial, and administrative software programs. COBOL, now 40 years old, continues to drive many large-scale financial systems today.

Over the next 40-plus years, Grace continued her work in computers in private industry and in public service as a naval commissioned officer, helping to set. the foundation for digital computing and programming as we know it. And while she called herself a mathematician, she spent her life demonstrating how mathematics principles could be applied to solve real-world problems, much like an engineer.

BLESSED WITH TREMENDOUS ENERGY, STRONG WILL, AND A FEISTY PERSONALITY, GRACE HOPPER CHANGED THE FACE OF COMPUTING. BUT THAT'S NOT WHAT MEANT THE MOST TO GRACE. SHE'S QUOTED AS SAYING, "IF YOU ASK ME WHAT ACCOMPLISHMENT I'M MOST PROUD OF, THE ANSWER WOULD BE ALL THE YOUNG PEOPLE I'VE TRAINED OVER THE YEARS; THAT'S MORE IMPORTANT THAN WRITING THE FIRST COMPILER."

Debugging the Bug

GRACE HOPPER and her team were trying to figure out why the Mark II computer wouldn't work. It turned out that a moth had gotten trapped inside, causing the hardware to short-circuit. When her boss asked what she was doing, she replied, "I'm debugging it."

The moth was removed and pasted into a logbook. The logbook entry reads, "First actual case of bug being found." The word "bug" thus referred to problems with the hardware.

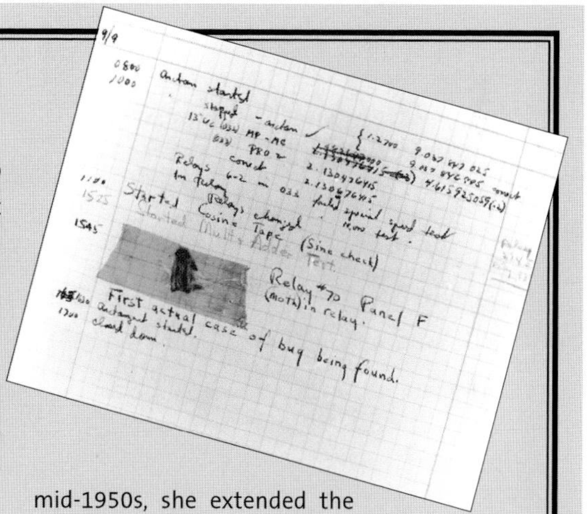

Grace, however, started using "bug" to describe an error in a computer program. In the mid-1950s, she extended the meaning of the term "debug" to include fixing programming errors.

The Grand Challenge

To pay off her college debt, in 1957 computer scientist and engineer **FRANCES "FRAN" ALLEN** (b.1932) took a temporary position with IBM, the largest computer company of the era. IBM was using supercomputers to help the military solve "grand challenge" problems—for example, forecasting weather and breaking secret codes.

Fran's job was to enable users of the fastest engineering scientific computer in the world to solve these large-scale problems. Fran built supercomputer compilers—the links between the programmer and the machine—to make it easier for people to use these highly complex computers.

Today, supercomputers are used to track global warming, and even model nuclear testing so scientists don't have to build real weapons to understand how they might work.

Upon retiring, Fran looked back fondly and said, "It was exciting to be part of the evolution of computing and have a ring-side seat as it changed the world."

The Stretch system was the world's first super-computer. The central processing unit alone spanned 33 feet, and was designed to be 100 times faster than any computing system of the time.

Radia Perlman enjoys playing piano as an accompaniment to her daughter, a violinist and opera singer. In one case, Radia wrote a poem about her famous "spanning tree" algorithm, widely deployed in networks today. Her son set it to music, and then she and her daughter performed this aria, called "Algorhyme," along with more traditional opera repertoire, at a local opera recital. "After all," she says, "every algorithm should have an algorhyme."

Directing Traffic

"I've always hated using computers," says computer scientist **RADIA PERLMAN** (b.1951), "so it became my goal to make them easier for people to use." Today, Radia is best known for developing fundamental tools—such as networks, protocols, and routing systems—that help support a constant stream of data.

A network is where computers talk to each other, the protocol defines how they talk to each other, and the routing commands tell the data which way to go. Radia's switching and routing protocols make it easy for everybody on a network to talk at the same time without concern for slowing down the system.

Radia says she's been able to come up with these ideas because of her unique—and very uncomputer-like thought processes.

"I was very intimidated by computing when I first started. But I liked to get things built, so I turned to engineering. Being able to think clearly and cleanly is very important. But it's also useful to see things from a different point of view. Some engineers just roll up their sleeves and start working. I like to step back and come up with a simple approach."

I NEED HELP!

Thinking about a vacation in Hawaii or maybe an adventure in Africa? One call or a quick website search can give you all the answers. This "automatic feedback" system is taken for granted today, thanks in part to the work of electrical engineer **LOUISE KIRKBRIDE** (b.1952).

Louise remembers, "It was frustrating to call a company and find that the answer I got depended on the person I talked to. I knew there had to be a way to automate these 'right' answers using the Internet."

She was right. By combining the Internet with a little programming, Louise was able to help businesses deliver automatic feedback online. Companies like American Airlines and Hilton Hotels use this system so that when we're ready to see the world, we'll know how to get there and where to stay.

"I like solving common problems that occur in everyday life," Louise says. "I figured if I have the problem, then other people probably do too."

IN HER SPARE TIME, LOUISE KIRKBRIDE LIKES TO SCULPT, DRAW, AND PAINT. "I TOOK ART CLASSES WITH MY ENGINEERING CLASSES IN COLLEGE AND HAVE CONTINUED TO EXPLORE DIFFERENT CREATIVE OPTIONS AS A HOBBY. THE TWO SKILLS GO TOGETHER NICELY. ENGINEERING IS SIMPLY PUTTING THOSE CREATIVE IDEAS INTO REAL-WORLD APPLICATION." AN ACTIVE TRAVELER AND PHOTOGRAPHER, LOUISE LOVES TO VISIT UNUSUAL PLACES AROUND THE WORLD. HERE SHE'S AT THE TOP OF A WATERFALL IN ICELAND.

HAPPILY WORKING IN THE UNIVERSITY ENVIRONMENT BOTH AS A TEACHER AND RESEARCHER, MARTHA SLOAN SAYS SHE GETS THE BEST OF BOTH WORLDS. "I'VE WORKED ON A RANGE OF PROJECTS FROM BETTER WAYS TO PROTECT DATA AND MESSAGES ON THE INTERNET TO INTEGRATING SOLAR PANELS INTO A WINDOW—ALL WHILE SITTING SIDE-BY-SIDE WITH THE NEXT GENERATION OF ENGINEERS AND SCIENTISTS, LIKE THIS YOUNG WOMAN."

A WOMAN IN CHARGE

With one eye on developing better forms of communication and the other eye on bringing members of a global engineering society together, **MARTHA SLOAN** (b.1939) became the first woman president of the largest professional organization in the world, the Institute of Electrical and Electronic Engineers (IEEE).

The executive board had never nominated a woman and therefore didn't put Martha's name on the ballot. So Martha gathered 6,000 signatures and became a write-in nominee. By election day, she had received over 25,000 votes, while her closest runner-up received just 15,000.

During her one-year tenure, Martha enjoyed bringing together engineers from other countries, particularly those living in Asia. "I traveled to Manila, Taipei, and Bangkok—to name just a few destinations—and worked with local engineers to share and develop information technology that will improve our world," comments Martha.

MARTHA SLOAN became the first woman president of the Institute of Electrical and Electronics Engineers (IEEE) in 1993. She was also the first woman president of the American Association of Engineering Societies in 1998.

She adds, "A lot of engineering is about talking to others and sharing ideas. It's what I like best."

Anytime, Anywhere

"Convergence" means coming together. In the telecommunications world, convergence means that all kinds of devices (such as phones, televisions, and computers) can handle all kinds of information and services (such as sound, images, and information).

It's real-time sharing between friends and colleagues. For example, people can now make phone calls on traditional land-line phones, cell-phones, cable lines, satellite, and the Internet. People can listen to radio stations on their computer. They can record movies on their cell-phones and send them to friends. The world is converging through communication. Some day, people, communities, and even nations will be able to find more common ground and the world will be more "connected" through telecommunications convergence.

"As an engineer, this time of development is so exciting. I have the ability to improve people's quality of life in very visible ways. The possibilities are enormous! It's engineers that will turn those possibilities into reality."

—Ruthie Lyle

Leveraging Broadband

One of electrical engineer **RUTHIE LYLE'S** (b.1969) projects at IBM is to help develop systems that speed up the transfer of information from the Internet to your computer. For instance, Ruthie has developed technologies that take advantage of broadband (meaning high-speed data transmission) wireless connections to the Internet.

Imagine being able to take as many pictures on a digital camera as you want. Or, download as many songs as you'd like without having to worry about the memory on your iPod®.

You'll be able to do this because the data will be stored somewhere on the Internet and not on the device itself. "The broadband wireless infrastructure is already in place," Ruthie says. "As data rates increase, new technologies will allow towns, cities, and even states to provide uninterrupted wireless Internet access."

In her spare time, Ruthie Lyle enjoys training and running in marathons. She's also studying Spanish to help develop a multicultural perspective. "I'm always looking to broaden my views and expand my ability to learn because that's how I will find new ways to help the world communicate."

All in the Family

In the early days, the Internet was not much more than an academic and government-sponsored experiment in communication systems. But, in the last 10 years, it's become an integral part of how we live our lives, a vehicle for building communities and bringing people together, as well as a way to interact with our environment.

JUDY ESTRIN (b.1954) looks ahead to see what's next. "I've always been interested in the Internet and its potential. Not what it does today, but what it can do tomorrow. We can then build the framework to make that possible."

The companies Judy co-founded are developing technologies that improve the performance of the Internet and make it easier for everyone to use. She says, "We're asking the Internet to do so much more than it was originally designed to do. We have to expand its existing framework to handle new demands."

Judy's sister, **DEBORAH ESTRIN** (b.1959), is finding ways to bring more information about the physical world to the Internet. Deborah develops wireless sensor networks—tiny embedded computers with specialized software and sensors that are often smaller than a cell phone—with tremendous power to gather data about our environment, take pictures, and then communicate the information back to us.

"As the technology matures, we might attach a sensor network in a forest canopy to study populations of endangered bird species, or embed these sensors in our water network, right into the pipelines that carry water to our homes. These sensors could immediately give us feedback about public health risks."

Deborah adds, "Networking research is about creating new communication systems—finding ways for individuals to coordinate and share information with one another—and, now with sensor networks, to let us understand and interact with the physical world in which we live."

Together, the Estrin family women have each changed the course of their respective engineering and science professions. From left to right: Daughter Judy used her engineering degree to become the CEO of Packet Design, LLC and chairman of Packet Design, Inc. Thelma Estrin—"Mom"—is an electrical engineer (see page 10). Oldest daughter Margo is a doctor of internal medicine. Deborah, the youngest, is professor of computer science at UCLA. And Dad? Well, he's an engineer, too.

"Until very recently, using the Internet for ordinary voice conversations was a pipe dream," Judy Estrin says. "Today the Internet has become almost as seamless a part of our everyday lives as the telephone."

Deborah Estrin frequently speaks to other researchers, engineers and the press about the intellectually-challenging problems that must be solved in order to make the emerging sensor technology a reality. She says, "This technology has so much to offer the world, and at the same time, it's a treasure trove of interesting design puzzles for us engineers."

Integrated Circuits

How can your cell-phone work better, smarter, faster? "Build better communication system hardware," says electrical engineer **RHONDA FRANKLIN DRAYTON** (b.1965).

Wireless, mobile, and satellite communication systems rely on high-speed circuits to move information from one point to another. Rhonda is discovering new ways to develop smaller, more efficient, and less expensive integration methods so that common communication devices such as cell-phones, or even your land-line telephone, can provide the clearest connection.

Rhonda says, "I'm like a detective. I don't actually design the individual pieces of the circuits. Instead, I look for ways to improve the connections between them. If I can make the links between these smoother, then cell-phone and land-line telephone service providers can support the increasing demand for instant anytime communication capabilities whether it's on a computer, a cell-phone, or maybe even through your television."

Called a silicon wafer, this high-performance piece of material is used to route signals in broadband high-speed communication systems that use silicon integrated circuit technology. This particular silicon wafer was designed, fabricated, and modeled by one of Rhonda Drayton's young graduate students at the University of Minnesota.

Seamless Mobility

Cell-phones are already in the "can't live without it" category for nearly 1.5 billion people. If chemical engineer **PADMASREE WARRIOR** (b.1969) has her way, these mobile devices will become even more indispensable. As Motorola's chief technology officer, Padmasree's global team of 25,000 scientists and engineers create technology breakthroughs for what she calls "seamless mobility."

"In today's world, people want to stay in touch no matter where they are—at home, work, in the car, or while they're out and about," says Padmasree. "We want to easily and securely access what we value, whether it's communication, information, or entertainment." Today's cell-phones are already so versatile that Padmasree refers to them as "devices-formerly-known-as cell-phones."

The fun is just beginning as researchers deliver cool technologies that allow you to instantaneously share images via your cellular call. Soon you'll be able to watch live TV on your cell-phone or even use your phone to save your favorite TV show to watch later. Now, that's a connection!

Stellar innovation is not new to Padmasree Warrior. A decade ago Padmasree guided the development of a breakthrough technology called RF LDMOS (laterally diffused metal oxide semiconductor). RF LDMOS has helped make today's faster networks and devices possible. RF LDMOS is now a fundamental part of the growing 3G (third-generation) wireless infrastructure market for Motorola, allowing the company to deliver phones with video playback, digital cameras, and more.

LIGHT UP THE FUTURE

In her senior year of high school, electrical engineer **KRISTINA JOHNSON** (b.1957) won first and second place in the International Science and Engineering Fair with a project she'd developed that used holography (three-dimensional images from lasers) to map the growth of fungus.

Kristina went on to build a research group that invented optical systems ranging from novel optical processors for detecting cervical cancer to projection televisions using liquid crystal on silicon (LCOS) display technology. Kristina invented the color management system used in these new televisions, which comprises one of her 44 patents. This work has been refined and commercialized by a company she started with her graduate and post-doctorate students.

Kristina is now the dean of engineering at Duke University but she's also keeping her engineering eye on scientific advancements in a relatively new area of study, called photonics. Just as electronics relies on power as its source, photonics relies on light, one of Kristina's favorite sources of study.

KRISTINA JOHNSON is the first woman dean of engineering at Duke University.

ABOVE: KRISTINA JOHNSON TALKS TO DUKE UNIVERSITY STUDENTS ABOUT THE CROSS-DISCIPLINARY ROLE OF ENGINEERS TODAY. "I TALK TO STUDENTS ABOUT SOLVING BIG PROBLEMS IN THE WORLD. THESE MIGHT RANGE FROM PRESERVATION AND RECLAMATION ACTIVITIES TO SAVING THE ENVIRONMENT TO RELIEVING PAIN THROUGH BIOMEDICAL DEVICES. WE CAN DO THIS WITH NEW INVENTIONS SUCH AS PHOTONICS."

LEFT: EVEN THE UNITED STATES CONGRESS RELIES ON KRISTINA'S EXPERTISE IN THE STUDY OF LIGHT TO HELP THEM DEFINE LEGISLATION THAT IMPROVES OUR COMMUNICATION METHODS.

"I think there's nothing better than inventing and building systems that make a difference to society."

—KRISTINA JOHNSON

"PHOTONIC TECHNOLOGY HAS REVOLUTIONIZED THE WAY WE TALK TO EACH OTHER," SAYS KRISTINA. HERE KRISTINA AND ONE OF HER STUDENTS WATCH AS A SINGLE OPTICAL FIBER (SILICON GLASS THAT TRANSMITS LASER LIGHT) CARRIES THE EQUIVALENT OF 300 MILLION SIMULTANEOUS TELEPHONE CALLS. SHE ADDS, "EVERY TIME ONE OF THESE DEVICES LIGHTS UP, IT SPARKS SOMETHING INSIDE ME. IT'S TREMENDOUSLY REWARDING TO WORK ON SOMETHING THAT WILL ENABLE US TO HANDLE 'ANYTIME-ANYWHERE' DEMAND THROUGH THE INTERNET—AND IT'S JUST AROUND THE CORNER."

WOMEN IN POWER

Electricity truly changed the world! It transformed all-day chores in homes and on farms into automated tasks. This gave more people, especially women, the chance to attend college and enter the workforce.

"Is it a fact—or have I dreamt it —that, by means of electricity, the world of matter has become a great nerve, vibrating thousands of miles in a breathless point of time."

NATHANIEL HAWTHORNE
AUTHOR (1804–1864)
FROM *THE HOUSE OF THE SEVEN GABLES*

Electricity also fueled an industrial economy beginning at the turn of the 20th century, when manufacturing plants—powered by electricity—started to mass-produce products.

It's easy to take electricity for granted. Just flick a switch, and on go the lights! In reality, electricity is the work of thousands of engineers who design machines to harness energy from coal, natural gas, oil, and radioactive materials. Other engineers develop networks to deliver the electricity from power plants to our homes, businesses, and industries.

The ever-increasing demand for electricity means that engineers must find ways to avoid black-outs and keep cities humming. Other engineers are dedicated to generating and distributing energy more efficiently, without harming the environment.

ABOVE: SOLAR POWER TOWERS, ALBUQUERQUE, NEW MEXICO.

RIGHT: GENERATORS.

Awesome Output

Isn't it amazing that when you stick a plug into an outlet, you get electricity? Where does that electricity come from, though? There are three main steps: generation, transmission, and distribution.

Electricity is generated at a power plant. A "turbine" (a wheel, blades, or fan) turns a shaft with a large magnet attached to it. The shaft and magnet are inside a coil of wire. As the magnet passes over the wire, it generates an electric current. The wire and magnet are called "generators."

The electricity is carried—or transmitted—through large high-voltage lines, sometimes over long distances. The electricity is then distributed to low- and medium-voltage users—that's us!

The first machines to generate electricity were developed in the 1880s. But there was plenty of room for improvement! Women engineers were instrumental in developing bigger, better, and more efficient power generators and turbines.

"Knowledge is power."

Sir Francis Bacon
Philosopher, Politician, and Author
(1561–1626)

Although we don't know for sure, Bertha Feicht most likely assisted her husband and brother behind the scenes in developing the Niagara Falls power plant, completed in 1895. The plant was a marvel in its day because it harnessed the power of the Niagara River (right) and provided power to the city of Buffalo, New York, some 22 miles away.

Background: Turbines.

Voice of Influence

As a child, mechanical engineer **BERTHA LAMME FEICHT** (1869–1943) and her brother, Benjamin, made their own toys and talked about starting their own toy company some day. The two did wind up at the same company, Westinghouse, where they worked with motors and generators—much larger "toys!"

Bertha attended engineering school to satisfy her curiosity about the way things worked. In 1893, she joined Westinghouse, where her brother had already become chief engineer. She and Benjamin worked in a design group that did early research in transmitting electricity. Bertha's future husband, engineer Russell Feicht, joined the group. But once he and Bertha married, company rules stated they could not work together.

Bertha retired in 1905. Many suspect she didn't stop working in engineering, though. She may have helped her brother and husband with their projects for Westinghouse. If so, she was a powerful influence!

To honor Bertha as the first women engineer at Westinghouse, the Westinghouse Educational Foundation, in conjunction with the Society of Women Engineers, established a scholarship in her name in 1973. The Westinghouse/Bertha Lamme Scholarship is awarded each year to a female college freshman studying electrical engineering.

Bertha Feicht's interest in math and her brother's love of his own engineering career enticed her to study engineering. Bertha is pictured here, around 1893, during her student days at Ohio State University.

Emma Barth helped Westinghouse customers buy the right power plant equipment to meet their needs. Here, she compares a manufactured turbine to the blueprints—plans and drawings—for the turbine.

In The Dark

Mechanical engineer **EMMA BARTH** (1912–1995) spent 33 years designing turbines and generators for Westinghouse. Specifically, she reviewed the needs of people buying her company's power plant equipment to make sure their purchases were a perfect match.

There was only one problem. Since many power plant managers would not allow women inside the plants, Emma could not see how the machinery, once installed, was working.

If someone reported a problem, a male engineer went to the plant, took photographs of the equipment, and described his observations to Emma so she could solve the problem. Imagine how difficult that was!

A few times, friends took her on plant tours. Or if her entire department went to a plant for a demonstration or photo, she could go, too. Emma would appreciate that, today, there are no such restrictions for women engineers.

Electric Expert

Electrical engineer **FLORENCE FOGLER BUCKLAND** (1898–1967) wielded great influence on steam turbines and other devices made by General Electric. Florence went to work at General Electric in 1921, straight out of college. She performed calculations to find ways to increase the power output in steam turbines.

Florence later became a heat transfer expert for General Electric. Other engineers sought her advice for improving everything from irons and kitchen ranges to locomotives and giant turbines. She had a knack for thinking of better ways to cool a motor or make a heating element smaller and lighter.

Florence also helped future generations of engineers by writing engineering reference books used by General Electric staff.

Above: The General Electric Company, Schenectady Works, 1907, where Florence Buckland worked.

Left: Florence (third from the left) worked for General Electric a few years, then went back to college. She earned a Master's degree in Electrical Engineering from Union College in 1925. Florence returned to General Electric and worked there until the 1960s, with a 16-year hiatus to raise her two children.

Hot Career in Heat

When mechanical engineer **NANCY DELOYE FITZROY** (b.1927) asked her father for a record player, he gave her one—unassembled. "My parents recognized my brain needed exercise," she says. With patience and care, Nancy built that record player. Later, she studied engineering to learn how more things worked.

> **NANCY FITZROY** was the first woman president (1986–1987) of the American Society of Mechanical Engineers.

After graduation, Nancy became a heat flow expert at General Electric. She wrote a book on the subject and delved into heat principles' many practical applications.

"I worked on everything from toasters and television tubes to submarines and satellites," she says. She designed better cooling systems for equipment, such as space satellites and aircraft jet engine afterburners. Nancy then created computer simulations to verify their higher operating efficiencies.

Although women engineers were rare early in Nancy's career, she says male engineers treated her well because they respected her work. Her husband of 53 years, and fellow engineer, Roland V. Fitzroy, Jr., was her mentor.

Besides working for General Electric, Nancy Fitzroy found time to fly her twin-engine airplane, *Nancy's Fancy,* and helicopters, too.

In Control

Power plant operators need to run their plants safely, reliably, and economically. To do that, careful monitoring, control, and coordination is required for all the plant's equipment: boilers, turbines, generators, and other machinery.

The flow of electricity from the plant must also be regulated. Additional instruments monitor power generation processes to be sure no problems, such as cooling water releases, damage the environment.

As power plants have grown in complexity, controls and instrumentation have become critical for proper plant operations. Mechanical engineer and control systems engineer **ADA I. PRESSMAN** (1927–2003) was an expert in power plant controls.

Working at Bechtel Corporation for most of her career, she saw control systems evolve from manual switches to complex automated systems. Ada was responsible for many breakthroughs leading to the development of these more precise and reliable sensors and controls.

During her career with Bechtel Engineers Power Division in Los Angeles, Ada Pressman managed 18 design teams for over 20 power plants, both nuclear and fossil fuel plants, throughout the world.

Among Ada Pressman's great achievements was the engineering of control systems for the San Onofre Nuclear Generating Station, which provides enough power to serve 2.75 million households in Southern California.

Americans use the equivalent output of five power plants every year to power electronics while they're turned off. So unplug while not in use!

The Juice Behind the Power

Most power plants in the United States use steam made in large boilers to drive the turbines. But what heats up the boiler to create steam?

Usually the heat comes from burning fossil fuels, such as coal, natural gas, and oil. Nuclear power plants use controlled nuclear reactions instead. In some parts of the country, steam comes directly from geothermal fields—places in the earth where water is heated naturally underground.

The fuel used depends on a region's available resources. For just over half of U.S. residents, that fuel is coal. Nuclear energy and natural gas each serve about one-fifth of the U.S. population. Hydroelectric power, generated at dams along large rivers and lakes, provides a small percentage of the country's power. Power plants fueled by oil and geothermal energy from natural steam fields round out the list.

Nicely Nuclear: Fission and Fusion

Nuclear power may sound mysterious. But it's just another way of making steam for power plants. So how does it work?

The key is nuclear fission. Think of a "fissure" as a "crack" or "split." Nuclear fission splits atoms to release heat. In a nuclear reactor, atoms of a nuclear fuel, usually uranium, are split to release heat. The heat powers boilers to make steam.

The opposite of nuclear fission is nuclear fusion. That's how the sun and stars get their energy. "Fusing" means "to combine." By combining atoms, mass is increased, which increases energy.

New Nukes

Nuclear engineer **KATHRYN A. MCCARTHY** (b.1961), a manager at the Idaho National Laboratory, is looking for new ways to harness nuclear power for energy.

Kathryn's area of expertise is nuclear fusion, and she has worked with a global team of experts to design a large-scale fusion reactor, the International Thermonuclear Experimental Reactor (ITER).

"My job was helping to prove the reactor would operate safely," says Kathryn. Construction on the ITER is expected to start soon.

Kathryn finds her natural empathy wins trust and eases the public's concerns about nuclear power. "Woman can have a big impact in this field," she says.

Above: As a manager at a top research facility, Kathryn McCarthy oversees work on both nuclear fission and nuclear fusion projects.

Right: The nuclear engineer describes her work as challenging and fun. She takes a lunchtime run every day to feel energized for busy afternoons.

Coal? Cool!

When electrical engineer **JILL S. TIETJEN** (b.1954) talks, people listen. She travels nationwide to give expert testimony on a number of power plant issues—from fuel costs to energy resources.

For example, Jill testified before Wyoming officials that two coal-fired power plants should be authorized for construction in Wyoming, home to many coal reserves. One, an 85-megawatt plant, now serves northeastern Wyoming communities, and construction started on the second plant in 2005.

"Electric utilities seek my opinion when they want to build a new power plant," says Jill. She has worked primarily with coal-fired plants, but understands other sources of energy, too. Jill helps utilities plan new investments. She develops detailed plans showing how a new power plant will work in concert with existing plants.

"I suggest ways the plant can operate as economically as possible and determine how much fuel it will require," she says. "When I look out from an airplane and see the lights on below, it gives me great pride to know that I help people have electricity to power their lives."

In addition to offering expert power plant advice, Jill Tietjen gives motivational speeches nationwide. She also co-authored the book, *Keys to Engineering Success* and *Setting the Record Straight: The History and Evolution of Women's Professional Achievement in Engineering*.

Invisible Energy: Natural Gas

Workers at the Katy Gas Plant near Houston, Texas, had never seen a woman at their facility until chemical engineer **CYNTHIA OLIVER COLEMAN** (b.1949) reported to work. "And it was the first time I'd been at a gas plant, too," she says.

Like most engineers, Cynthia had developed strong problem-solving skills in college. But she would learn much on the job. In a short time, she became an expert in natural gas plant operations.

Later, Cynthia became an "energy detective." She surveyed natural gas fields and used computer programs to determine how much natural gas the field might yield and how to recover the most fuel possible.

As the first African-American woman chemical engineering graduate from the University of Houston, Cynthia felt honored to share her story here. "I carried out my mission to encourage others—especially minorities—to study engineering," she says.

Cynthia Coleman spent her entire 32-year career with ExxonMobil so she could remain in Houston with her family. Her daughter is now a medical doctor in Houston.

Power to the People

Electricity is still not commonplace in some Asian, African, and South American countries. But people in the United States and other developed nations worldwide can thank the engineers who pioneered transmission systems every time they flip a light switch or power on a computer.

Women engineers were involved in early power transmission systems. Today, they work for utility companies as design engineers and managers, making sure power is always available for our increasingly electronic world.

Rural Voltage

Electricity became commonplace starting in the 1890s. In cities, that is. Although nearly 90 percent of urban dwellers had electricity by the 1930s, only ten percent of rural residents did.

The Rural Electric Administration, established in 1935, took on the job of bringing electricity to outlying parts of the country. Power transmission and distribution grids further expanded to provide inexpensive electric lighting and power to farms.

Electrical engineer **HILDA COUNTS EDGECOMB** (1893–1989) helped the Rural Electric Administration in its efforts to bring electricity to America's farmers. She reviewed engineering plans for transmission lines and substations and made sure the facilities designed would meet their goal of delivering power to remote areas.

Bringing electricity to farms meant that tasks could be automated. With fewer hands needed, children raised on farms could pursue higher education.

Human Calculator

A hundred years ago, an orphaned girl was encouraged to save her inheritance. Electrical engineer **EDITH CLARKE** (1883–1959) went against this advice by spending her nest egg on a college education. She studied math and astronomy so she could have interesting work.

Edith taught math and then became a "calculator" for American Telephone and Telegraph (AT&T) in 1912. In the early 20th century, the only calculators were the human kind. People were hired to do the mathematical calculations needed to design electrical equipment.

At AT&T, Edith learned about engineering, then went back to college for an electrical engineering degree. After graduating, however, no one would hire her as an engineer. She was told that women were only suited to be "calculators," or "computors," as they were also called.

When General Electric finally gave her an engineering job, her assignment was to find ways to transmit electricity over great distances with as little loss of energy as possible.

Edith developed transmission line charts that simplified the calculations and reduced the time needed to solve problems in design and operation of electrical power systems. Edith also patented a method to regulate voltage on power transmission lines so the lines could operate near maximum power limits and provide electricity to more people.

By simplifying power transmission calculations, Edith Clarke reduced the need for hand calculations, typically performed by women, and opened up the once male-dominated electrical engineering field.

Bright Lights, Big City

Over 3,000 miles of distribution circuits serve Seattle City Light's 380,000-plus customers. As director of central electrical services, electrical engineer **ELISABETH "BETTY" A. TOBIN**, makes sure the power is always on in downtown Seattle's underground network.

Betty was the first woman electrical engineer hired by Seattle City Light. She worked in several departments including a stint designing substations and facilities to generate and transmit electricity. "I liked doing all aspects of the work, not specializing, so I learned a lot," she says.

Now in an engineering management position, Betty takes a "big picture" outlook encompassing political, legal, and personnel issues. "I enjoy advocating for my employees, and getting them what they need to do their jobs well," she says. She keeps her design skills sharp through active involvement in IEEE, an electrical engineering society.

Betty Tobin credits her engineering background with making the move into utility management possible. "The type of job I do is not something you can just fall into without qualifications," she says.

Lessening the Impact

Meeting the public's need for power is not the only concern for power companies. Utilities must be sensitive to the impact that power plant operations may have on the local environment and to unique regional needs.

Women engineers working in power help preserve habitats, the places where native plants and animals, including endangered species, live. Others teach utility customers how to conserve energy to stretch limited energy reserves and reduce the need to generate power.

To move away from the use of fossil fuels, women engineers are working in the public and private sector to generate energy from clean, eternally-renewable sources: the sun, wind, and water.

Solar thermal systems can provide electrical power for industrial, commercial, and residential customers. Solar power is a sustainable, environmentally-friendly option, but its higher operating costs, compared with traditional energy sources, must also be considered when evaluating energy alternatives.

Habitat Enhancements

In one of her first jobs for a power plant near the Atlantic coast, industrial engineer **TERESA A. HELMLINGER** (b.1953) designed a slide to help tiny shrimp and other marine life return to the ecosystem rather than get pulled into a pump drawing water for the plant. Her inspiration for the low-friction device? Fiberglass swimming pool slides!

In another project, Terri and her team delivered power to an uninhabited island so the marines at a nearby military base could do training exercises. "The only catch was that part of the island was the habitat of an endangered woodpecker," says Terri. While most of the power lines were buried underground, Teresa and her team used overhead power lines through the habitat. "That way, the woodpeckers had new, tall power poles for perching," she says.

In 2003, **TERESA HELMLINGER** became the first woman president of the National Society of Professional Engineers.

Terri advanced to managerial roles with Carolina Power & Light. She and project teams found ways to save money and increase revenues for the utility.

After a career with Carolina Power & Light, Terri Helmlinger became the director of a North Carolina State University group offering technical assistance to state businesses and industries.

GREEN GOALS

The term "sustainability" refers to a way of life in which economies, people, and nature flourish. The word was not used in engineering school when mechanical engineer **JANICE PETERSON** (b.1950) got her degree. Now, her work is all about sustainability. And she loves it!

She meets with Portland General Electric's business customers in their workplaces to suggest ways to use less electricity. "I'm constantly away from the office," says Janice. "But that's OK!"

Janice offers a complete, resource-saving package by advising business owners in water use, waste disposal, and ways to minimize trash going to landfills as well. For companies that want to build an energy-efficient building from the ground up, Janice works with building owners, architects, and construction contractors during the design and building stages.

JANICE PETERSON LIKES BEING IN THE FIELD, MEETING PEOPLE, AND SEEING PLACES THE PUBLIC RARELY GETS TO GO—DATA CENTERS, MECHANICAL ROOMS, ROOFTOPS. "I GET A BEHIND-THE-SCENES LOOK AT A LOT OF COMPANIES," SHE SAYS.

SOUTHWESTERN SUN

Just a few years ago, the Navajo Nation in parts of Arizona, New Mexico, Utah, and Colorado lived too far from established power grids to get electricity in their homes. "Bringing the energy to them would have cost $27,000 per mile, far more than tribe members could afford," says Native American civil engineer **SANDRA K. BEGAY-CAMPBELL** (b.1963). "Many lived over 10 miles from the grid!"

Sandra helped the Navajo Tribal Utility Authority place solar units on individual homes to generate electricity. The cost for homeowners is $75 a month. A much better deal!

Sandra has helped hundreds of tribe members get electricity. "I'm the helpful hand that comes with federal funding," Sandra says, referring to a Department of Energy Tribal Energy Program in which grant recipients receive her technical assistance as part of a package.

In addition to solar energy, Sandra looks into other alternative energy sources such as wind power, geothermal power, and biomass (trash and industrial wastes burned for energy).

AT LEFT, SANDRA BEGAY-CAMPBELL POSES WITH A SOLAR CAR. ABOVE, SHE STANDS IN FRONT OF A SOLAR SYSTEM PROVIDING POWER ON ONE OF HER PROJECTS.

She proudly passes on the engineering tradition by mentoring teens. In addition, three of her cousins are studying to be engineers, and a niece now in high school plans to join the profession, too!

"I get to work with my own native people. I give them a new way to think about having electricity. It's very nurturing."

—SANDRA BEGAY-CAMPBELL

High-Flying Women

The dream of flying like a bird has drawn the curiosity, creativity, and daring of women and men around the world. People get a thrill when they envision soaring above the treetops with a bird's eye view of the splendor below!

Ruth Bancroft Law (above) was the first woman enlisted as an Army aviator on June 30, 1917.

They love the feel of wind whipping past them and the crisp cool air. They see complete freedom from the confines of a terrestrial life. No walking, only flying.

A 1944 Boeing Stearman PT-17 Bi-plane.

Even before Wilbur Wright's first powered flight in 1903, many flight visionaries tried to accomplish the "impossible" with everything from strapped-on wings to engine-powered hydrogen balloons.

Turning their dreams of flight into reality became the passion of some of the world's most ingenious woman inventors and engineers. From boxy bi-planes to agile fighting machines, women bright with thrill and promise have helped take the aerospace industry to new heights.

"Reach high, for stars

lie hidden in your soul."

PAMELA VAULL STARR

POET, ARTIST, AND SONGWRITER (1909–1993)

BOEING 757 ASSEMBLY LINE.

F/A-22 RAPTOR.

Flight Takes Off

Many of the first airplanes were built—and rebuilt—by experimentation and observation, as opposed to any scientific analysis. Getting an airplane off the ground was one thing, but making airplanes practical and useful for the general public was another. To do so safely and successfully required an understanding of why airplanes flew or "aerodynamics."

> *"I wanted a life which would never be boring—a life in which new things would always occur."*
>
> —Irmgard Flugge-Lotz

The first planes were often built of lightweight woods such as spruce or bamboo, covered with fabric. During World War I, engineers moved to stronger—although more costly—aluminum alloys and high-alloy steel materials.

The first airplane engines generated about 12 horsepower—less than the power generated by today's riding lawnmowers. By comparison, a 747 jet generates about 500,000 horsepower! These three elements—aerodynamics, structural materials, and engines—defined the basics of flight.

In these early times, men were as new to this exciting field as women, so the opportunities for women to contribute as engineers were perhaps more open than other fields.

The Wright brothers' test gliders launched in 1902 at Kill Devil Hills, North Carolina, were precursors to the first powered airplane.

How Airplanes Fly

For an airplane to fly, it must have lift. It's not the engines that provide lift—it's the wings. Early aerodynamicists understood this important concept and worked to analyze the flow of air above and below the wings while an airplane is in motion.

IRMGARD FLUGGE-LOTZ (1903–1974) used her study of fluid mechanics (how fluids move over objects) and math to demonstrate how much air flows over which parts of the wing at any given time during flight. In the 1930s, Irmgard began working with Ludwig Prandtl, considered the father of modern fluid mechanics.

She made an immediate impact on her mentor by developing an easy way to measure how lift is distributed from wingtip to wingtip—a solution that Prandtl and many others had been unable to solve for more than a dozen years! Applying her math skills, Irmgard was able to solve this lift distribution problem and develop a practical method for improving airplane wings.

During the rest of her career, mostly at Stanford University, Irmgard went on to make advancements in automatic control theory that would eventually lead to the development of the autopilot and other advanced flight control systems used on both commercial and military aircraft today.

Despite the Nazi policy denying women leadership positions, Irmgard Flugge-Lotz acted as the head of the aerodynamic theory department within the top German aeronautical research institute in the 1930s, though she had no official appointment. She was just 31 years old, and her boss, the Director of the Institute, recognized her genius and provided her the space to apply her mind successfully.

From Models to Machines

A stenographer by day, airplane builder by night, for **E. LILLIAN TODD**, the chance to build a flying machine occupied her mind and heart. By just experimenting and talking to other airplane enthusiasts, she was able to design and build a bi-plane. This plane was exhibited in Madison Square Garden in 1906, though it didn't fly. It would take another four years for her to find an engine to meet her design criteria.

While she continued to improve her design and search for an engine, she also helped teach others about her new passion. In 1908, she helped establish and lead the Junior Aero Club of the United States. The members of this club—all teenage boys except Lillian—made model planes and attempted to fly them at the regular meetings.

Lillian Todd watched with trepidation and excitement as the bi-plane she designed and built tolled down New York's Garden City Aviation Field on November 8, 1910. She had asked an "experienced" pilot to navigate her airplane down the field. As she watched, it flew 20 feet, turned around and came back to where it started, much to the thrill of Lillian and her friends who watched. Now she'd have to do it herself!

Flying with Safety

In the early days, pilots simply looked out the window to make sure nothing was in the way—and then they took off! As air travel became more common in the 1920s, the need for safety followed—particularly after a couple of dramatic and deadly mid-air collisions.

The U.S. government took on responsibility for defining flight regulations, licensing pilots and aircraft, and constructing airfields. The Civil Aeronautics Administration (now, the Federal Aviation Administration—FAA) was formed in 1940 to emphasize safety for pilot and passengers. Because women made up a small yet strong group of pilots during this era, they also made sure their voices were heard when it came to flying with safety.

"Being a woman should not deter you from being an engineer . . . if women really want to, they can do anything. I see unlimited opportunities."

—Katharine Stinson

A Passion for Flight

AMELIA EARHART (1897–1937), one of the best known aviators of the 1920s and 1930s, inspired more women to fly and to learn the basics of flight than perhaps any other person. In 1929, she established the "Ninety-Nines," an organization for women pilots who came together to share experiences and advance aviation.

Amelia Earhart was fiercely competitive when it came to flying. Just two years after taking lessons, she broke flying records and the women's record for high-altitude flight. She was also the first woman to fly across the Atlantic with two male pilots. Not long afterward, she became the first woman to fly solo across the Atlantic.

Despite her short life, Amelia's quest for knowledge and her passion to do the unusual has inspired young women around the world.

Safety in Numbers

As a young girl growing up in rural North Carolina, **KATHARINE STINSON** (1917–2001) dreamed of learning to fly airplanes. Her first chance came at age 10, when aviation pioneer Eddie Stinson (no relation) took her up in his enclosed cockpit plane. She was hooked.

Katharine Stinson met her heroine, world-famous aviator Amelia Earhart, when she was 15. It was 1932 when Earhart landed in Raleigh, North Carolina, to service her plane. Stinson worked as a mechanic's helper at the airport to earn money to help pay for flying lessons. When Stinson told Earhart that she was learning to fly, Earhart urged her to pursue a career in aeronautical engineering.

With a bachelor's degree in mechanical engineering with an aeronautics option, Katharine joined the Civil Aeronautics Administration, specializing in aircraft safety or "air worthiness." After developing standards for airplanes, the FAA then tested and monitored the airplanes to make sure they were safe. She helped develop the standards for supersonic transport, which were used to build the Concorde jet.

Katharine received the FAA's Sustained Superior Performance Award in 1961. From 1964 until her retirement in 1973, Katharine was the technical assistant in the FAA's Engineering and Manufacturing Division, second in command to the chief, the highest grade for a woman in the FAA.

Katharine Stinson's first application to the college of engineering at North Carolina State University was rejected, so she completed two years of coursework at Meredith College in one year, returned to North Carolina State, and graduated with distinction in 1941. That year, she was one of only five women in the country to graduate with an engineering degree!

Katharine knew she wanted to be an engineer at the age of 15 after meeting her heroine, Amelia Earhart. At left, Katharine holds the picture of Amelia and herself taken in 1932.

A Look Inside

In the 1940s, engineers wanted to better test airplane parts to make sure they were flight-ready for many flight hours and through many weather conditions. Mechanical engineer and metallurgist **REBECCA SPARLING** (1910–1996) pioneered the field of "nondestructive testing" to support these studies.

Nondestructive testing relies on sound waves to "hear" if a metal or metal object has flaws, without having to break the object apart. It's not much different than thumping on a watermelon to see if it's ripe, except instead of your knuckles you use an electronic device to generate sound waves. When the wave pattern varied, Rebecca could tell the location and the size of the flaw.

While at Northrop Aircraft in 1948, Rebecca's group developed an even more advanced method, called the immersed ultrasonic inspection method, which helped when evaluating parts such as those used inside engines. Nondestructive testing methods—both underwater and in dry conditions—are still used today to evaluate material strength and consistency.

Below: Even the tiniest flaw could lead to major disaster inside a jet engine. That's why Rebecca Sparling looked carefully at the inside of an engine nozzle using one of her newly designed nondestructive testing probes. As the waves bounced off the engine nozzles, the sounds "told" Rebecca where there were possible problems.

Right: The DC-3, first flown in 1935, included sleeping berths and an in-flight kitchen. These airplanes paved the way for the modern American travel industry and became a mainstay of air transportation until the 1970s.

Imagine flying an airplane that is cruising at a little over 2,000 miles per hour! The SR-71 Blackbird, built in the 1960s, is the fastest piloted airplane ever built.

"I'm a success at failures. Whenever you work with failure analysis you have to think on all levels. You can't limit yourself to things you've learned in the past."

—Esther Williams

Reliable Materials

Understanding why airplane structures crack or joints fail under certain conditions leads to stronger, more reliable designs. With a background in metallurgy and an engineer's mind for problem-solving, **ESTHER HAWLEY WILLIAMS** (1914–2000) pushed the status quo of her day, seeking materials that could be used for aircraft able to fly faster than 2,000 miles per hour and higher than 85,000 feet.

The heat generated while maintaining these kinds of speeds would cause materials such as aluminum to fail. Steel was an option, though it was heavy and therefore not optimal for airplane design.

While at Lockheed Missiles and Space as a materials reliability engineer, Esther suggested—and then proved—that a newly available material called titanium might work. Esther demonstrated that titanium holds its shape at these tremendous speeds and is light enough for aircraft structures. Her work gave birth to the SR-71 Blackbird, the world's fastest and highest-flying airplane of the time.

Esther Williams had a knack for figuring out what caused an airplane to crash. For much of her career, her job was to determine what happened to an airplane by putting together the pieces left over at a crash site. As an expert on materials and how they behave in various conditions, she was able to figure out where the failure occurred, and then why. Esther used this expertise to support design and development of everything from modern day aircraft to the Hubble Space Telescope.

The War Years

The early years of flight overlapped with World War I and World War II. As men went off to war, the aeronautical industry turned to women to fill roles as engineers and technicians. However, many technical schools still didn't accept women.

The aircraft industry couldn't wait! Instead, from 1941 to 1944, specialized courses were offered to train women as engineers and for other technical jobs. Women saw this as an unprecedented opportunity to enter a new career and earn significantly higher pay.

These industry-based programs also helped change the attitude of traditional engineering schools toward women. More and more women were given the opportunity to study in their engineering field of interest.

Elsie MacGill managed the production of the Hawker Hurricane fighter plane. In only four years, she and her team constructed 1,400 airplanes. Amazing! In memory of her work, a comic book was published honoring her famous nickname, "Queen of the Hurricanes."

Queen of the Hurricanes

For **ELIZABETH "ELSIE" MURIEL GREGORY MACGILL** (1905–1980), engineering in the 1930s and 1940s held a world of possibilities, full of excitement, energy, and creativity. Despite being crippled with polio while in college and using crutches to help her walk, Elsie earned her degree in aeronautical engineering and went on to work for the Canadian Car and Foundry Company to build airplanes.

During World War II, Elsie was chief engineer of the classic Hawker Hurricane fighter plane, the Canadian Royal Air Force's first fighter monoplane with an enclosed cockpit, retractable wheels, and the ability to exceed 300 miles per hour in level flight. She then adapted it for cold weather. If not for these airplanes and Elsie's efforts, the battles for British air superiority in 1940 might have taken a far different turn.

Elsie was also responsible for the manufacturing of the Hawker Hurricane fighter plane. At the peak of production, Elsie supervised a staff of 4,500.

Elite Few in Aviation

In the United States, the development of airplanes for commercial and military aviation began in 1915 with the creation of the National Advisory Committee for Aeronautics (NACA). The NACA's mission was to improve all facets of aviation—from aerodynamic principles to aircraft manufacturing to licensing.

The NACA engineers were an elite few who helped establish America's dominance in aviation research and construction, particularly during and after World War II. For electrical engineer **KITTY O'BRIEN-JOYNER** (b.1916), it was the opportunity of a lifetime to explore an exciting new world.

Kitty joined the NACA in 1939. Beginning in the early 1960s until she retired, Kitty was the head of the cost-estimating group for facilities construction projects. This included managing multiple wind tunnels that were built to better understand supersonic flight and new airfoil designs. Kitty also helped develop modern-day aircraft standards.

At a time when women were rolling up their sleeves to handle male-dominated jobs like setting rivets or wielding welding irons, Kitty O'Brien-Joyner used her electrical engineering expertise to help reshape the Langley Aeronautical Laboratory for a new era of flight.

A Radio Rebel

As early aviators began to fly more, they also looked for better ways to navigate. Initially pilots relied on landmarks and other visible locations on the ground such as beacons. Of course, these landmarks were only helpful during daylight and good weather!

Radio communication and other electronic navigation tools revolutionized aircraft navigation, allowing pilots to fly anytime in all weather conditions and still track their location and destination. For **ALICE MORGAN MARTIN** (d.1998), developing this technology was an early passion.

Against her parents' wishes, she spent more than a decade studying mechanical and electrical engineering in night school while working at other jobs during the day. One of her first jobs was with Bendix Aviation Corporation, where she directed a team of engineers in the design of radio control units for commercial and military aircraft.

Over several decades, Alice worked with major airlines and aircraft manufacturers to improve aircraft navigation, communication, and control units that made a difference in airplanes and even in space.

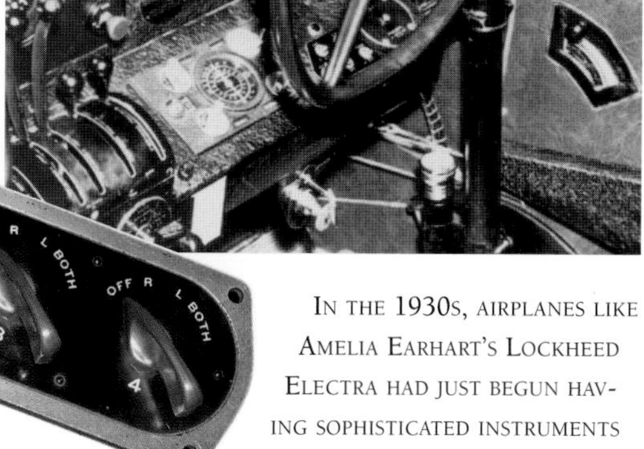

In the 1930s, airplanes like Amelia Earhart's Lockheed Electra had just begun having sophisticated instruments installed, such as longer-range radios and gauges that allowed a pilot to fly without seeing the ground below. That's critical when flying high over large expanses of water.

The Modern Machine

As stick controls and small engines gave way to sophisticated electronic systems and the modern jet, people had the opportunity to dream bigger, travel further, and explore the many diverse cultures that make up our world.

In the late 1940s, after World War II, a new era of air transportation introduced commercial airplanes that seated up to 100 people. By the late 1950s, over one million people flew to Europe from the United States. Today, airlines transport more than 600 million people every year within the continental United States alone!

Throughout this time, women engineers have improved passenger comfort, designed lighter aircraft, developed more fuel-efficient engines, and worked to minimize environmental concerns for people around the world.

"Of all the women I know who have studied engineering, few of us have worked directly in the field that we graduated in. However, the foundation of a solid engineering experience at the college level opened doors for all of us to move into areas we'd never have considered."

—Elizabeth Plunkett

Quiet Please!

So as not to annoy passengers, aircraft are thoroughly analyzed and tested to minimize engine or airflow noises, and structural vibrations. Vibrations can also lead to structural deterioration. Engineers perform these studies in a test stand or wind tunnel, as well as on completed aircraft still in the factory and then in flight.

As a senior program engineer in Boeing's Structures/Noise Technology division, mechanical engineer **ELIZABETH I. PLUNKETT** (b.1925) and her team took responsibility for improving safety, passenger comfort, and aircraft durability for many of the industry's notable passenger planes.

Today, commercial air travel is quieter and structurally safer because of Elizabeth's work on commercial airplanes from the 727 to the 767.

Elizabeth Plunkett called her quest for an engineering degree "the most successful disaster I've ever been through." After several years of studying engineering and dealing with the negative attitude of mostly male teachers and students, Elizabeth left school and went to work for Boeing where she was able to apply her ability to design, analyze, and test airplanes. It wasn't long before Elizabeth became a senior program engineer with responsibilities for improving many of today's most successful commercial airplanes.

Cabin Comfort

For Boeing mechanical engineer **JEANNE YU** (b.1962), cabin comfort is a top priority. As the technology leader for the cabin environment design team on the 787 Dreamliner, Jeanne wants to make the cabin more people-friendly.

She says, "I find out where the technology is ready and then bring it onto the airplane. It's an ongoing creative process for our diverse team to work together to develop the best features which will add quality, value, and comfort. These range from comfortable seats to improved environmental control systems that better manage air quality."

As part of her technology leader role, Jeanne is actively involved in the Federal Aviation Administration (FAA) Airliner Cabin Environment Research team conducting research and development on the healthfulness of the cabin environment for passengers and crew.

Jeanne Yu is working with the FAA Airliner Cabin Environment Research team on finding ways to protect passengers against pesticide or chemical contamination, to monitor ozone exposure, and then to put in place technologies to monitor and even resolve problems.

FUTURE AIRPLANES BIG AND SMALL

Who knows what the shape of tomorrow's airplane will be? Can we create a "green" airplane that doesn't require jet fuel? Can planes be more people-friendly?

Aerospace engineer **PEG CURTIN** (b.1954) and mechanical engineer **JULIE CHAPMAN** (b.1980) say "yes" to all these questions. Peg and Julie spend much of their day testing airplanes in wind tunnels.

Peg uses the wind tunnel testing to make new Boeing commercial airplanes—like the 787 Dreamliner—more efficient, quieter, and easier to operate. "By making small structural changes like reshaping the wings, less fuel is required to operate the plane, and the cabin is quieter for passengers," says Peg.

Julie is doing research on airplanes small enough to fit into backpacks! These robotic planes carry technology, not people. They fly in places too dangerous for manned flight, to provide an eye in the sky.

"Because they're smaller and less expensive to construct, these small airplanes also give us a chance to test new airplane building concepts. Designers massage and change their airplane components to achieve more speed, more fuel efficiency, lighter weight—or all three. Then we test them in the wind tunnel to see if the design performs as planned, and look for the causes of any unexpected results."

Says Julie, "I never thought I'd be on the ground floor of future aircraft design—but here I am."

LEFT: WHEN JULIE CHAPMAN WAS YOUNG, SHE LISTENED TO HER FATHER, ALSO AN ENGINEER, TALK ABOUT THE BASICS OF A CAR GENERATOR AND OTHER MACHINES. SHE SAYS, "I ALWAYS THOUGHT I COULD MAKE IT BETTER." TODAY, JULIE DOES JUST THAT, THOUGH HER MACHINE OF CHOICE IS AN AIRPLANE. IN HER JOB WITH THE AIR FORCE, SHE FLIES MODELS OF AIRPLANES AND PARTS IN WIND TUNNELS TO IMPROVE THEIR CHARACTERISTICS AND CAPABILITIES.

RIGHT: STANDING IN THE WIND TUNNEL AT THE NASA LANGLEY NATIONAL TRANSONIC FACILITY, PEG CURTIN SAYS, "HERE IS WHERE I WORK TO IMPROVE AERODYNAMIC DESIGN. I ALSO COMPARE THE WIND TUNNEL DATA TO COMPUTER MODELS OF THE CONFIGURATIONS WE TEST, WHICH HELP IMPROVE THE COMPUTER MODELS TO AID IN FUTURE DESIGNS."

CAROLYN MERCER USES HER PATENTED OPTICAL SYSTEM TO MEASURE THE EFFECTS OF VARIOUS ATMOSPHERIC AND FLIGHT CONDITIONS OF ICE FORMATION ON AIRCRAFT, ENABLING MORE EFFECTIVE DE-ICING SYSTEMS.

ENGINES AND THE ENVIRONMENT

Aerospace engineer **CAROLYN MERCER** (b.1960) looks at commercial aviation from a slightly different angle. The NASA-sponsored Ultra-Efficient Engine Technology (UEET) program team focuses on developing new turbine engine propulsion technologies for commercial aircraft that are quieter, safer, and more environmentally friendly.

As the technology manager for the UEET program, Carolyn says, "My role is to develop technologies today that will be used 10 to 15 years from now on jet engines for aircraft. I target specific engine design-related problems, host competitions for new ideas, select new tasks, and make sure the work gets done effectively and then gets transferred into the hands of those in private and public industry who can use the new technologies."

RUNNING CLEAN

In recent years, the aviation industry has become increasingly aware of its impact on the environment. This has led to jet engines that are much cleaner and quieter than they were just 10 years ago. However, international regulatory agencies are demanding ever-lower noise and emission levels.

Mechanical engineer **JEANNE ROSARIO** (b.1952) and her team at General Electric Aircraft Engines certify, test, and evaluate a new generation of engine technologies for commercial aircraft. These engines are the core of many familiar airplanes from the 737 to the new 787. "I know that with every test or certification, I'm doing my part to support people and the environment," she says.

Jeanne adds that from her earliest memories, she wanted to know how things worked. "I'd crawl under a draw-bridge just to see the mechanism begin to turn," she recalls.

JEANNE ROSARIO STANDS BENEATH THE WORLD'S LARGEST, MOST POWERFUL TURBOFAN ENGINE—AN ENGINE SHE HELPED DEVELOP. THIS ENGINE NOW POWERS THE BOEING 777 AIRPLANES.

A Big Wind Blows

The world's largest wind tunnel is big enough for a full-size passenger jet to fit inside! It's located at NASA's Ames Research Center in Mountain View, California. The six fans that blow air in this massive building each have a diameter equal to the height of a four-story building. The winds produced by the fans, combined with the narrowing tunnel, can speed air flow to more than 250 miles per hour. NASA also has a wind tunnel where models as small as a match can be tested.

Certain Ceramics: A Day in the Life of Katherine Faber

Developing cutting-edge technology, teaching and mentoring students, and researching new areas are all integral parts of being a college professor. There's never a dull moment for ceramic engineer Katherine T. Faber (b.1953). Her workday at Northwestern University is filled with interesting people and fascinating ideas. And she lives and works in Chicago, one of the country's most exciting metropolitan areas!

Katherine is an expert in ceramic coatings, which are applied wherever heat is a factor. While metals in engines soften or melt at temperatures over about 2,000°F, metals coated with ceramic can withstand another 300 or more degrees. Ceramic coatings are used in the aircraft, aerospace, automotive, and power industries. They enable engines to increase fuel efficiency by running at higher temperatures.

Katherine believes there are many discoveries still to be made in her field. "Research rarely has a definite ending. With each new development comes new questions. That's exciting, because there are always opportunities to go in a new direction and explore new things."

Chicago, viewed from the Northwestern University Campus where Katherine works.

"I found I was more attracted to engineering than pure science because engineering had a problem-solving focus. I am a problem-solver."

—Katherine Faber

Beyond the classroom and lab, Katherine leads students in extracurriculars. Here she's advising a group of graduate students involved in a local after-school science program for sixth to eighth graders. "I really enjoy getting to know students better. I can see their leadership skills develop, too. That is particularly satisfying," she says.

All of Katherine's research is conducted in teams with students and colleagues. "I enjoy the creativity of a team environment, especially when the group brainstorms ideas for improvements"

With graduate student Ariel Knowles, Katherine is studying ancient jade objects housed at the Art Institute of Chicago. Their work helps art curators and engineers understand the chemical makeup of the pieces and solve old mysteries, such as why a 3,000-year-old jade figure has a curious shiny surface.

Katherine's day also presents many opportunities to confer with other colleagues on exciting new developments in research and technology. Here, she and Professor Lincoln Lauhon take a look at the Ph.D. thesis prepared by one of Katherine's former students.

"Living in a large city means there's always something to do, whether it's educational, cultural, or sports-related. There's an energy associated with the city that's contagious," Katherine says. She and her family enjoy local sports and performances of the Chicago Shakespeare and Steppenwolf Theaters. On vacations, they seek out other big city sights, such as San Francisco's crooked Lombard Street.

Breaking Barriers

For the aviation world, breaking through the sound barrier means traveling at supersonic speed—faster than the speed of sound! For woman pilots, breaking barriers can mean more than just traveling fast. It takes know-how, too.

Airplanes are extremely complex machines, and a pilot who is also an engineer has a huge advantage in flying them with skill! To be a good pilot, it's simply critical to understand the principles of flight and the equipment within an airplane. Engineering is, thus, an inseparable bond that links women to their planes.

Speed Demon

To this day, aviator **JACQUELINE "JACKIE" COCHRAN** (d.1980) holds more speed and altitude records than anyone else—man or woman. She's shown here standing on the wing of her Sabrejet F-86 fighter jet, which she flew in 1953 to become the first woman to break the sound barrier.

An orphan, Jackie grew up in poverty in rural Florida. As a teen-ager, she became a beautician at a local hairdresser's and in 1929 moved to New York to start her own line of cosmetics. She learned to fly in order to cover the territory necessary to sustain a cosmetics business. And fly she did!

This F/A-18 Hornet creates a shock wave—visible as a large cloud of condensation formed by cooling of the surrounding air—as it breaks the sound barrier.

FLYING FIGHTERS

For systems engineer **MARY "MISSY" CUMMINGS** (b.1966), the U.S. Secretary of Defense's 1993 decision to open the door for women to participate in military combat—and thus fly a new generation of super-sleek fighter airplanes—changed her world. Prior to 1993, female pilots and other female aircrew were restricted to non-combat programs. As one of the Navy's early female fighter pilots, Missy relied on her engineering knowledge to excel.

While Missy admits that life as a pioneering aviator was not easy in the Navy's super-male-dominated environment, it was her love of flying that kept her going.

After 10 years in the Navy, Missy moved to the academic world as a professor of aeronautics. She's currently studying how humans interact with technology.

"Pilots of today's aircraft must not only understand basic aerodynamic principles but also complex operation of automated flight control systems. Automation can be helpful, but when it fails it can also be deadly. It's critical that we develop strategies and systems that allow humans and automation to work together."

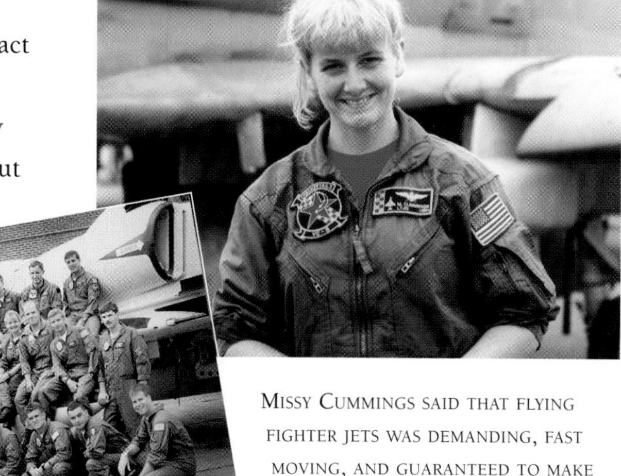

MISSY CUMMINGS SAID THAT FLYING FIGHTER JETS WAS DEMANDING, FAST MOVING, AND GUARANTEED TO MAKE YOUR HEART RACE. "LOW-LEVEL MISSIONS ARE THE BEST—YOU FLY OVER 500 MILES PER HOUR, NO MORE THAN 100 FEET ABOVE THE GROUND, WATCH THE CLOCK, WATCH FOR OBSTACLES, AND ARRIVE ON TARGET INSIDE A FIVE SECOND WINDOW!"

SOARING WITH THE EAGLES

"Flying wasn't something that I grew up wanting to do," said **STAYCE HARRIS** (b.1959). "I was planning to be an engineer. Turns out the two fit together. Flying is an analytical skill—basically problem solving—just like engineering."

Stayce's opportunity to fly came in her first year of college when her Air Force ROTC instructor suggested she apply for pilot training. She said she'd give it a try. Not long after, Stayce began her Air Force career, soaring through the skies in the agile supersonic, two-seater T-38 training jet. "It was like strapping a rocket on your body and taking off," she recalls.

Today, Stayce, a colonel in the Air Force Reserve, says her career has helped her look at the world from a different perspective. "I've helped deliver humanitarian support to nations around the world—from medical relief to the construction of schools and homes devastated by human or natural disaster. Engineering helped me find what I love and a place to make a difference."

STAYCE HARRIS TAKES CHARGE OF THE AIR FORCE RESERVE'S 459 AIR REFUELING WING. SHE'S RESPONSIBLE FOR OVER 1,300 PILOTS, FLIGHT CREWS, AND MAINTENANCE AND MISSION SUPPORT PERSONNEL INCLUDING MEDICAL TEAMS AND CIVIL ENGINEERS—AND GETS TO FLY PRETTY MUCH WHENEVER SHE WANTS.

Marta Bohn-Meyer studied engineering so that she could talk to pilots—and fly advanced airplanes. Marta says this about her success and the thrill of flying airplanes like the Mach 3 SR-71, "I had the 4 R's of success—the right place, right time, right qualifications, and the right enthusiasm."

Left: In her private life, Marta was also an aviator and a world-class aerobatic pilot. She and her husband built this aerobatic plane called the Giles G300. With this plane, Marta competed as part of the U.S. Unlimited Aerobatic Team in world competitions.

Planes With Pilots . . . and Without

At age 14, aeronautical engineer **MARTA BOHN-MEYER** (1957–2005) learned to fly in a small two-seat airplane. Twenty years later, she became the flight test engineer for the super-fast, super-sleek SR-71 Blackbird. Marta and her team used this airplane as a testbed to fly scientific experiments. These included using an ultraviolet astronomy camera to measure ultraviolet light at very high altitudes and even a unique type of rocket engine that was flown on the back of the SR-71 to test capabilities. Marta said, "I'm really the link between the airplane behavior while flying and the engineers and researchers who are still on the ground."

As chief engineer for Dryden, Marta applied that knowledge to build UAVs—unmanned aerial vehicles. UAVs are aircraft that can take off, fly, and land, but are electronically "piloted" by people on the ground.

"UAVs can do things in the air that we can't do. For instance, if we could park UAVs at 60,000 or 70,000 feet above the Earth, we could use them to transmit better cell phone connections. Firemen could use them to spot fires before they get out of hand. Today, the military uses them to take pictures and gather information about locations where terrorists are believed to live. It is a very exciting time."

Virtual Airport

What will the future airport look like? **NANCY DORIGHI** (b.1952) has a pretty good idea. As the manager of FutureFlight Central, part of NASA Ames Research Center, she and those on her team operate one of the world's premier airport operations and planning design studios.

Their facility is able to replicate a 360-degree high-fidelity visual simulation of any airport in the world. FutureFlight Central can provide a functionally accurate, physical and software replication of any airport's current or future tower or operations center.

The two-story facility offers a full-scale real-time simulation of an airport, where controllers, pilots, and airport personnel participate to optimize expansion plans and operating procedures, and evaluate new technologies. Nancy and her group provide a detailed and highly realistic 3-D airport database model displayed on twelve projection screens to provide a 360-degree out-the-window view of the airport.

Airport research and operations staff can work with NASA experts using FutureFlight Central to plan new runway configurations, test new ground traffic and tower communication procedures, and validate air traffic planning simulations based on airport and airline planning tools.

FutureFlight in space? In addition to planning for more efficient, higher-capacity airports of the future, Nancy Dorighi is thinking about how these simulations might support the study of Mars. She hopes these advanced simulations will provide the live video background for information sent from rovers as they drive around the Red Planet.

How many more planes can an airport handle with a new runway? Will it be safe? Airport planners work with NASA experts to answer those questions. In a virtual airport, airport planners are able to compare alternative designs that just exist on paper, and air traffic controllers fine-tune how they will guide the airplanes on the ground. Planning for the future of air travel means squeezing out every ounce of efficiency and, at the same time, keeping it safe for the flying public.

Stars in Space

Is there life in outer space? How are planets created—or destroyed? Does space offer answers to problems here on Earth? Can we save our own planet, which has been used for thousands of years by billions of people? These are the questions that have sparked generations of space research.

For some, it's the thrill of the unknown, the need to know if there are other

Left: Space shuttle launch. Above: Earth viewed from moon.

beings that exist as we do. For others, it's about understanding who we are and where we came from.

It's about finding cures to disease and hunger here on Earth. It's about finding ways to preserve what's left of the Earth's natural resources, the all-important ozone layer, and our environment.

Above: Andromeda Galaxy.

Left: Satellite data showing the Antarctic ozone hole in 2001. Dark blue colors correspond to the thinnest ozone, while light blue, green, and yellow pixels indicate progressively thicker ozone.

THE RACE FOR SPACE

The real possibility for space exploration began soon after World War II with the development of high-powered rockets. Thus began a gripping political and technological competition between the United States and the former Soviet Union.

Who would be the first to build a rocket that could launch a satellite or spacecraft into orbit? Who would be the first to land on the moon?

Once engineers and scientists had figured out how to get a satellite into orbit and then bring it back safely, it wasn't long before we launched humans. The Soviets were the first to put a robotic spacecraft on the moon, dogs in space, a man in space, a woman in space, and to walk in space.

But on July 16, 1969, with his famous words, "That's one small step for man, one giant leap for mankind," Neil Armstrong and his crew on Apollo 11 achieved the dream: a lunar landing.

Pups Fly First

It was actually two female dogs, Belka and Strelka, who made the first roundtrip into space and back in 1960 on one of the Soviet Union's Sputnik launches.

BACKGROUND: AERIAL VIEW OF APOLLO 11 BEFORE LAUNCH.

COMMAND AND CONTROL

Every satellite, spacecraft, missile, and airplane contains an extraordinary number of controls, switches, and sensors to monitor operations. These small but mighty pieces of equipment alert the onboard systems—pilot or machine—when problems occur. Adjustments can be made and the problems fixed.

Chemical engineer **BEATRICE HICKS** (1919–1979) pioneered the design, development, and manufacture of these unique devices. From environmental sensors to molecular scanners and her patented gas density switch, Beatrice's work allowed satellites and, later, people to maneuver in space with confidence.

In 1950, **BEATRICE HICKS** became the first president of the Society of Women Engineers.

Beatrice is one of the few women to head a technology company during the early days of the space race. She took over her family's business, Newark Controls Company, as president and director of engineering.

DESPITE A SIGNIFICANT LACK OF ENCOURAGEMENT FROM FAMILY MEMBERS OR EDUCATORS, BEATRICE HICKS BELIEVED THAT WOMEN ENGINEERS HAD THE ABILITY TO CHANGE THE WORLD—AND PROVED IT BY PIONEERING SYSTEMS THAT HAVE HELPED SHAPE SPACE TRAVEL OVER THE LAST 50 YEARS.

BRING 'EM HOME, BOBBIE!

For **BARBARA "BOBBIE" CRAWFORD JOHNSON** (1925–2005), the Apollo program was a time of great excitement and even greater anxiety. As a Rockwell engineer, Bobbie made sure that the Apollo spacecraft made pin-point landings—at exit and reentry.

The Apollo spacecraft, developed during the 1950s, was the vehicle for the U.S. mission to the moon. Bobbie was one of the first engineers to work on the Apollo program. She designed the re-entry guidance instrument, a back-up system called an Entry Monitor System (EMS) that helps the astronauts make a safe return to earth.

"I worked on the Apollo proposal design and development. And every time we'd have a flight, one of the astronauts would look at me and say, 'Is the EMS going to work, Bobbie?' I think the lunar landing was the *coup de grace* for me. It was great!"

BARBARA JOHNSON WAS A BRILLIANT SPACE-AGE ENGINEER, AND, ACCORDING TO ONE OF THE 100 MALE ENGINEERS AND TECHNICIANS WORKING UNDER HER AT THE TIME, "FUN TO WORK WITH. [SHE] HAS A GOOD SENSE OF HUMOR AND ALWAYS A SMILE EVEN WHEN THE WORKLOAD IS UNBEARABLE."

ROCKET WOMAN

Launching people and objects into space requires millions of pounds of thrust—and that's just to get off the ground! Rocket engines are at the heart of space travel, particularly the high-performance liquid hydrogen fuel rocket engines, like the RL10, initially developed in the early 1960s to take Apollo astronauts to the moon.

These amazing engines are also at the heart of one woman's career as an engineer. Just 18 years old, **SUSAN H. SKEMP** (b.1944) took a job as an engineering aid for Pratt & Whitney, an aerospace and aircraft engine manufacturer. "It was supposed to help me pay for my night classes. Turns out it was the start of a very exciting career."

Sue used this introduction into high-performance engine development as a stepping stone to a mechanical engineering degree, and was present when the RL10 was dedicated as an American Society of Mechanical Engineers historical engineering landmark. Upon graduation, she became a performance systems analysis engineer for a new generation of fighter aircraft engines for the Navy and Air Force. These engines are now used to power the F-15, F-16,

SUE SKEMP HAS PUT INTO OPERATION SOME OF TODAY'S MOST ADVANCED AIRPLANES AND SPACECRAFT. SHE RECENTLY MOVED INTO THE OFFICE OF SCIENCE AND TECHNOLOGY POLICY, IN THE EXECUTIVE OFFICE OF THE PRESIDENT OF THE UNITED STATES, TO USE HER ENGINEERING SKILLS AND BACKGROUND TO HELP ESTABLISH PUBLIC POLICY.

and F-22 fighters.

Sue says, "Working with engines in the lab and on the test stand has been so exciting. It's a hands-on chance to really watch the engine perform and then find ways to make it quieter, generate lower emissions, and extend capability using better materials and new designs."

Sue recommends finding "a way to shape what you love with your unique aptitudes so that it fits your needs and interests." Sue's daughter took her advice. She earned a degree in music engineering technology so that she could combine her love of music with a technical skill set. She launched her career working with a company to develop a new era of high-definition TV.

There and Back Again

Landing on the moon effectively ended the space race—but served only to fuel human interest in further space exploration! The U.S. set its sights on developing a true space vehicle, one that could go back and forth from space many times.

Thus began the Space Shuttle program—actually called the Space Transportation System (STS). The shuttle began to be designed in the early 1960s, but didn't roll out onto the Cape Canaveral (Florida) launch pad until April 1979.

The Space Shuttle represents the work of thousands of engineers. The shuttle is used for many different projects including the launch of new earth-monitoring systems, telescopes, and planetary probes, extensive biomedical experiments, and visits to space-based stations.

In 1998, Bonnie Dunbar managed all payload activities on the STS-89 *Endeavour* mission including 23 technology and science experiments. The crew of this mission also transferred more than 9,000 pounds of scientific equipment, logistical hardware, and water from *Endeavour* to the Russian space station, *Mir*. Today, Bonnie is the head of the Museum of Flight in Seattle, Washington.

Woman on a Mission

In third grade, **BONNIE DUNBAR** (b.1949) watched the Russian Sputnik satellite make its way through space—and she was fascinated! She wanted to fly, but she wasn't sure how.

When a college professor showed her a picture of the Space Shuttle, then under design and construction, she knew the answer. The professor had a research grant to work on the all-important ceramic shuttle tiles, the thermal protection system that allows the shuttle to safely—and repeatedly—re-enter the earth's atmosphere.

NASA announced in 1977 that it would be selecting women for the Space Shuttle program. Bonnie, now a ceramic/mechanical/biomedical engineer, was first in line!

Bonnie became an astronaut in 1981. She's flown five missions and has been heavily involved in the study of microgravity (where there is little or no gravity).

Bonnie says, "It's because of engineering, the systematic thought process, that I've been able to achieve the things that seemed impossible to others. It's given me a chance to solve world problems, change the environment around us, and make life better for so many."

In Safe Hands

"I like to make things work—whether it's flying a helicopter or operating a 50-foot robotic arm while floating in space," explains astronaut and safety engineer, **NANCY CURRIE** (b.1958). Nancy is NASA's expert on robotic system development and application.

But inside NASA, Nancy is best known for her unwavering commitment to safety: "I wanted to fly for the military ever since I read a story about a helicopter pilot who evacuated injured soldiers out of combat zones. I joined the Army and went on to fly helicopters, where I eventually became an instructor.

"It was one training day that changed my life. At the last minute, I was pulled from my normal flight team to go up with a visiting flight instructor. While we were in the air, the other helicopter had a mechanical problem and crashed. All of my fellow pilots were killed while I looked on in horror.

"At that very moment, at age 22, I changed my career plans. I went back to school to become an engineer. A safety engineer, to be exact, because safety engineering would allow me to influence how an airplane is built, how it's flown, and how it's maintained—all with an eye on safety for the pilot and crew."

Nancy says that the overall strategy in safety engineering is threefold: eliminate as much risk of malfunction as possible through good engineering design, provide warning or safety devices for impending failure, and teach pilots how to address a problem if there is a failure.

"In 1987, the Army sent me to NASA as a flight simulation engineer for the Shuttle Training Aircraft. I qualified as an astronaut in 1990.

"As a flight engineer, my focus on safety engineering became even more critical, particularly after the Shuttle disaster in 1986. The Shuttle is a very complex system. At take-off and landing (the two most critical times), I make sure that if a computer failure occurs, I understand what that means to other systems so that the crew can respond appropriately and very quickly.

"Being an engineer means knowing how the entire system works. You have to know how failure in one system can affect other systems. We have flight crew astronauts that don't have engineering backgrounds—but it's a steep learning curve."

Part of the 2002 Shuttle mission was to repair and service the Hubble Space Telescope. Nancy Currie used the Shuttle's 50-foot robot arm to catch the Telescope, help astronauts during the work, and then boost the telescope into a higher orbit once work was complete.

"Don't ever let anyone look you in the eye and say you can't do this. If you have the skills and the desire to succeed, you can achieve anything. There are no barriers"

—Nancy Currie

NASA Women Engineer-Astronauts

Note: The "STS" and numbers following astronauts' names are the shuttle flights on which they have flown.

Mary Cleave, environmental engineer: STS-61B, 30

Catherine "Cady" Coleman, polymer science and engineering: STS-73, 93

Nancy Currie, safety, industrial engineer: STS-57, 70, 88, 109

N. Jan Davis, mechanical engineer: STS-47, 60, 85

Bonnie Dunbar, ceramic, biomedical, mechanical engineer: STS-61A, 32, 50, 71, 89

Susan Helms, aerospace engineer: STS-54, 64, 78, 101, 102, 105

Marsha Ivins, aerospace engineer: STS-32, 46, 62, 81, 98

Mae Jemison, chemical engineer: STS-47

Tamara "Tammy" Jernigan, engineering science: STS-40, 52, 67, 80, 96

Susan Kilrain, aerospace engineer: STS-83, 94

Wendy Lawrence, ocean engineer: STS-67, 86, 91, 114

Sandra Magnus, electrical, material science engineer: STS-112

Karen Nyberg, mechanical engineer

Ellen Ochoa, electrical engineer: STS-56, 66, 96, 110

Heidemarie Stefanyshyn-Piper, mechanical engineer

Nicole Stott, aeronautical engineer, engineering management

Janice Voss, electrical engineer: STS-57, 63, 83, 94, 99

Mary Ellen Weber, chemical engineer: STS-70, 101

Joan Higginbotham, electrical engineer

Kathryn "Kay" Hire, engineering resources management, space technology: STS-90

K. Megan McArthur, aerospace engineer

Lisa Nowak, aerospace engineer

Sunita Williams, engineering management

Stephanie Wilson, engineering science, aerospace engineer

In Memory of Those Who Dreamed

Since the dawn of the space age, the world has been treated to some unbelievable accomplishments—and some devastating losses. NASA's two most recent losses include the Space Shuttle *Challenger,* which blew up 73 seconds after launch in 1986, and the Space Shuttle *Columbia,* which came apart on re-entry into Earth's atmosphere in 2003 after a two-week mission in space.

NASA and the world lost seven remarkable people on each of these missions, including two extraordinary women engineers. These individuals died pursuing a dream, searching for answers, and improving life for those of us on Earth.

KALPANA CHAWLA (1961–2003). A licensed commercial pilot and flight instructor who enjoyed flying aerobatics, Kalpana loved to fly. In January 2003, Kalpana flew on her second mission on the Space Shuttle *Columbia.* She and the crew conducted over 80 experiments.

On her first of five Space Shuttle missions, Judith Resnik worked with the crew to conduct a crystal growth experiment and photography experiments using the IMAX motion picture camera among numerous projects, while orbiting the Earth 96 times.

JUDITH RESNIK (1949–1986) flew her first shuttle mission on *Discovery* in 1984. She had said, "I think astronauts probably have the best job in the world."

Judith's father said, "She had the brains of a scientist and the soul of a poet."

As a mission specialist, Kalpana Chawla hung on in the weightless space environment while she studied test results to better understand astronaut health and safety, as well as the effects of microgravity on humans and objects during her 16-day flight.

Survive and Thrive

For those of us on Earth, space is a pretty hostile environment. It's basically a vacuum where there is no air, pressure, or gravity. There's no protection from the sun's radiation and heat.

That's why engineers design space vehicles, space stations, and spacesuits so that those in space can live and work in an Earth-like environment. To learn what it takes to live in space, you must, well, . . . live *in space.*

Thus the development of space-based "homes," like the International Space Station, began soon after people were launched into space. Little more than laboratories, these stations are designed to help us better understand what's needed for humans to survive in space.

Using this protected environment, engineers must develop ways to supply the basic necessities of life—air, food, and water—for long stays in space.

Background: Solar flares send out intense radiation that can be harmful to astronauts, equipment, and those of us here on Earth.

Where Food Takes Thought

In the early days of space flight, the menu was pretty limited. Most food was stored in tubes or freeze-dried and then rehydrated. Today, even though most food is still pre-cooked, dehydrated, and reconstituted, astronauts can occasionally eat fresh fruit and vegetables from unmanned resupply rocket launches.

In the future, space travel to the moon or Mars will require enough food to last several weeks or years. The additional weight and storage requirements exceed the current shuttle's capacity.

That's why biomedical engineer **DEBORAH WELLS** (b.1965) is trying to find ways to grow food while in space. Seeds for fast-growing foods like onions, radishes, lettuce, and strawberries require much less room than the food itself.

"This is a true engineering challenge," explains Deborah. "We've done plenty of plant growth experiments in space, so that we know space travel changes the growth. On earth, we'd use fluorescent light to stimulate growth, but that's far too energy-intensive for space. There's still a great deal to learn."

Fast-growing, nutritious food is a top requirement for astronauts taking a long flight to Mars. Deborah Wells is looking for ways to grow food in space.

Easing Water Limits

Water is so costly to launch into space that each crewmember on the International Space Station is allowed only 4.4 gallons per day for everything—cleaning and drinking. The average American uses 90 gallons per day on Earth!

Mechanical engineer **CINDY HUTCHENS** (b.1957) has devoted her career at NASA to developing space-based water and wastewater recycling technologies. Cindy and her team have made a vapor compression distillation system that turns crewmembers' urine and wastewater into clean water for drinking, cooking, and cleaning. This urine processor has already been tested in space, and a full-scale system will be sent to the Space Station very soon.

"Each crewperson uses 473 gallons per year. That's not much water, considering it includes all the water used to cook, flush, drink, and bathe," Cindy says.

"We recover 739 gallons of urine per year. This is treated and added to the water available to a six-person crew. The entire water recovery system produces 2,840 gallons annually."

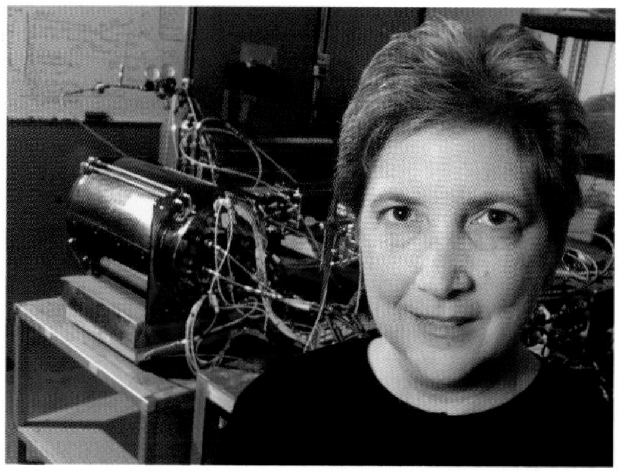

With help from Cindy Hutchens' space-based water recycling system, the International Space Station (below) crew will soon have nearly twice as much water available for drinking, cooking, and bathing.

STANDING IN THE ARIZONA DESERT, AMY ROSS (LOWER RIGHT) AND HER TEAM LOOK ON AS ONE OF NASA'S TEST SUBJECTS DEMONSTRATES HOW TO USE A THE ROBOT REMOTE CONTROL TO ACQUIRE SOIL SAMPLES. THE ASTRONAUT MUST CLIMB INTO AND OUT OF THE LARGE ROVER VEHICLE, THEN BEND DOWN AND WORK WITH THE SMALL MOBILE SOIL SAMPLER. SEEMS LIKE AN EASY TASK UNLESS YOU'RE WEARING AN OUTFIT THAT WEIGHS 250 POUNDS AND FEELS AS IF YOU'RE WEARING A LARGE MARSHMALLOW.

LIKE HAND IN GLOVE

Not long ago, each set of spacesuit gloves was custom-designed for each astronaut, and fitted to ensure maximum mobility and security while working in space. It was a job that mechanical engineer **AMY ROSS** (b.1971) took very seriously, and perhaps more personally than others.

Amy designed the gloves worn by her father, Jerry Ross, when he flew in the 1998 *Endeavour* mission. Her dad spent over 21 hours working in space to assemble parts of the International Space Station. His gloves took a full 18 months to build, weighed about three pounds apiece, and were able to withstand temperatures from -250°F to 250°F.

Amy says her group doesn't make very many custom glove sets anymore. There are enough different sizes that most crewmembers can wear an existing pair. That's okay with Amy. There's lots more work to be done.

The temperature difference alone on the moon is daunting: the average daytime temperature on the surface is 224°F while the nighttime average is -243°F. Add to this the chemical, radioactive, and biological conditions on a planet such as Mars, and these systems—from gloves to boots to bodysuits—must be pretty resilient!

"It requires an engineer to create practical systems that allow people to exist in these unfriendly environments," says Amy. "For instance, I have to understand what kind of soil is found on Mars, and how its chemical makeup will affect the spacesuit materials.

"My dream is to see a Mars mission during my career—I know we can do it and I plan to be part of it."

She adds that engineering is a professional career that continually gives her a feeling of self-worth and pride, a chance to affect how the world works. "There's always a challenge. I'm always growing. Engineering has given me the confidence to move forward, and helped me learn how to learn."

"HIGH" FASHION

The modern spacesuit is much more comfortable than the 1950s-style spacesuit, which was so stiff it was difficult for the astronaut to sit. The newer designs improve the range of motion, though it's still difficult to bend down and pick up a pencil!

Tomorrow's spacesuits must be more flexible so that if future space explorers want to hike the dry and dusty hills and craters of Mars, they can do so with ease. Along with greater mobility and flexibility, the suits themselves must weigh a lot less.

Aerospace biomedical engineer **DAVA NEWMAN** (b.1964) explains, "Astronauts need not only the ability to walk and run, but the flexibility to rappel off a cliff or crawl through a cave." Her innovative Bio-Suit could well be the answer.

"I'm designing a suit for the extreme explorer," Dava says. "It will fit like a second skin so that movement is as natural as it is here on Earth." This body-shaped suit will also include embedded communications equipment, biosensors, computers, and other gear that might be needed to climb around unfamiliar territory.

The spacesuit of tomorrow uses advanced materials and technologies to fit like a second skin. Dava Newman is looking into cutting-edge fabric such as advanced "muscle wire" technology and electrospinlacing (a process that spins fabric using electric charge). Dava says creating this suit requires the innovative ideas of a multi-disciplinary team. "I work with professional designers, architects, advanced materials experts, other engineers, specialists in life-support systems, and many others. As an engineer, I bring it all together."

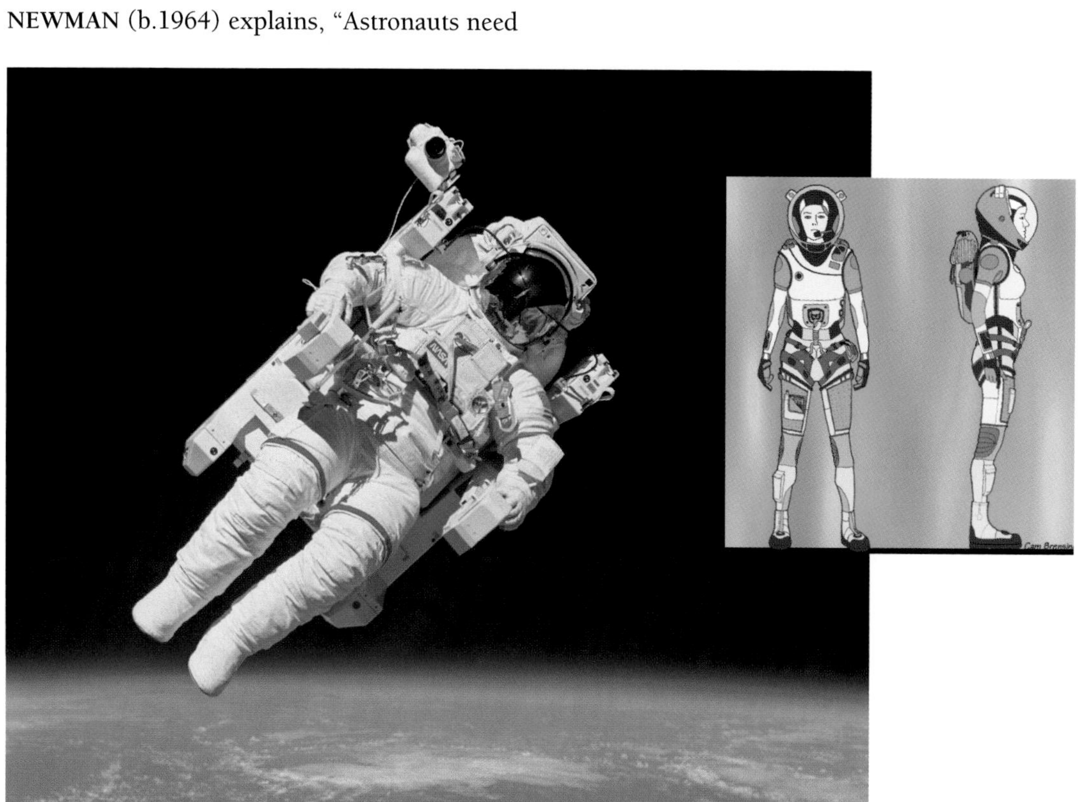

"This is engineering at its best.

I bring people and ideas together

and turn them into reality."

—DAVA NEWMAN

THE SEARCH FOR ANSWERS

Is there life on other planets? How did the universe begin? Are there other galaxies like ours? These are big questions that have no clear answers . . . yet.

To see distant planets, watch the formation of stars, or study another universe, you must get beyond the earth's protective atmosphere deep into the darkness of space. The Hubble Space Telescope, launched into Earth's orbit in 1990, helps scientists and astronomers study the history of our universe. Hubble is able to see objects that are over 12 billion light years away—that's equivalent to 70,540,000,000,000,000,000,000 miles!

Actually visiting these other worlds is another step altogether—and we've done that too! Over the last 40 years, spacecraft and probes have orbited and landed on Venus and Mars, explored the sun, tracked comets and asteroids, and moved within close-range of Mercury, Jupiter, Saturn, Uranus, and Neptune.

From the rovers roving around Mars to the Cassini-Huygens *orbiter and space probe landing on Saturn's moon, Titan, the search for life on other worlds continues to awe and amaze those of us here on Earth.*

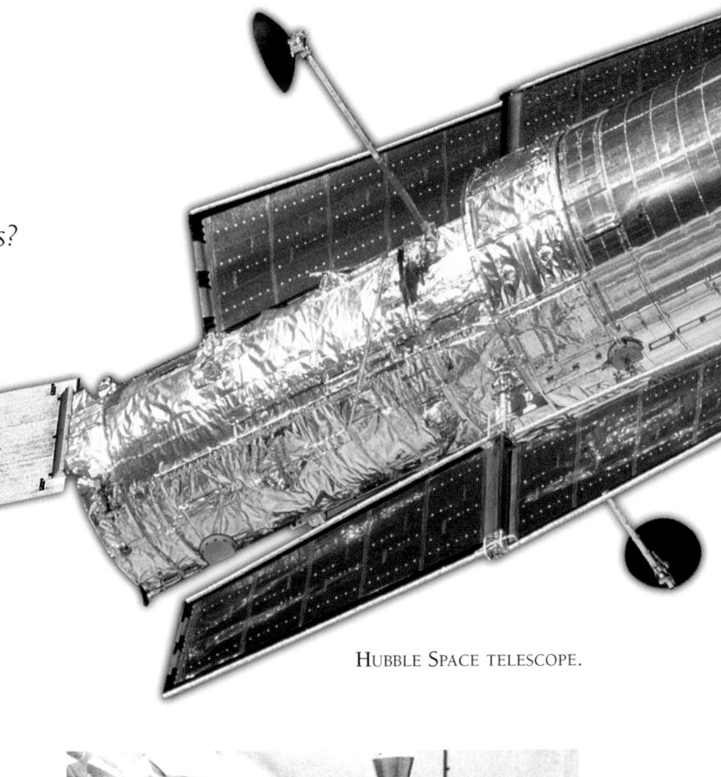

HUBBLE SPACE TELESCOPE.

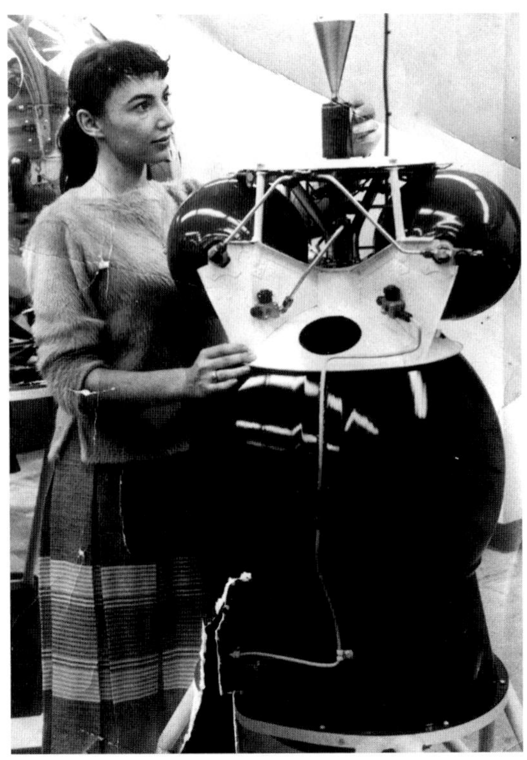

IN THE 1960S, IN HER EARLY YEARS AS AN ENGINEER, JUDY COHEN HELPED DESIGN THE *ATLAS-ABLE* LUNAR PROBE, ONE OF THE SPACE INDUSTRY'S FIRST ATTEMPTS TO EXPLORE LUNAR SURFACES. NOW SHE WRITES CHILDREN'S BOOKS, INCLUDING *YOU CAN BE A WOMAN ENGINEER*.

A LOOK BEYOND

In the 1970s and 1980s, NASA envisioned a telescope that could peer into space and help uncover the mysteries of the universe: a giant earth-orbiting telescope that could be turned in any direction to study everything from black holes to the birth and death of stars, even comets crashing into other planets or the evolution of entire galaxies!

NASA began developing the Hubble Space Telescope, and electrical engineer **JUDITH "JUDY" LOVE COHEN** (b.1933) wanted in on the ground floor. It was 1980 and Judy already had over 20 years of experience under her belt—in communication systems for satellites, missiles, and other defense-related projects.

She became the systems engineer—the one that does the overall design—for the science operations ground system (SOGS). SOGS is what enables astronomers and scientists from around the world to schedule observation time, process digital film, track celestial objects, and much more—in short, to see the universe!

Judy says, "I grew from a young girl that had never even heard of a female engineer to become the leader of a team of engineers on the design of the Hubble Space Telescope. That was an incredible journey!"

THE *CASSINI-HUYGENS* SPACECRAFT HAD TWO KEY ELEMENTS, THE *CASSINI* ORBITER AND THE *HUYGENS* PROBE. JULIE WEBSTER KEEPS THE *CASSINI* SAFE AND HEALTHY WHILE IT SPORTS AROUND SATURN AND ITS MOONS ALMOST A BILLION MILES AWAY. "IT'S PRETTY AMAZING. WE ANTICIPATE THE NEXT FOUR YEARS TO BE NOTHING SHORT OF SPECTACULAR IN TERMS OF NEW SCIENCE DISCOVERIES ON SATURN."

FLYING RINGS 'ROUND SATURN

When the *Cassini-Huygens* spacecraft, launched in October 1997, finally reached Saturn in June 2004, nobody was more excited than systems engineer **JULIE WEBSTER** (b.1953). "I've been with this thing since it was a hunk of aluminum. Getting *Cassini* into orbit around Saturn and releasing the *Huygens* probe on its way to Titan were the culmination of ten years of truly incredible work."

Julie made sure that the spacecraft's capabilities would meet the scientists' needs for gathering information. "There are always limited engineering resources, like data-gathering and storage and how fast the spacecraft can turn," she says.

In December 2004, *Cassini* released the *Huygens* probe into the atmosphere of Titan, Saturn's largest moon. Titan is the only moon in the solar system with its own atmosphere. The data will be radioed to the *Cassini* orbiter, which will then relay the data to Earth.

The three-foot-long, 703-pound *Huygens* probe has already sent information about Titan's atmosphere, spectral data, and much more. Meanwhile, the *Cassini* orbiter continues to fly within Saturn's orbit to study the planet, the rings, and the many moons. Over four years, *Cassini* will complete 74 orbits of the ringed planet.

Julie adds, "We tried to accommodate the science needs without taking unnecessary engineering risks—and based on results so far, it's working!"

"IT LOOKS A BIT LIKE A LEGO PROJECT OR A FUTURISTIC PLANT STAND," SAYS DONNA SHIRLEY, STANDING NEXT TO THE *SOJOURNER* MICRO-ROVER JUST PRIOR TO ITS 1997 TRIP TO MARS. MADE MOSTLY OF ALUMINUM WITH TITANIUM WHEELS, THIS TINY DATA-GATHERING DYNAMO TRAVELED 119 MILLION MILES THROUGH SPACE TO ROLL ACROSS THE ROCKY SURFACE OF MARS FOR 83 DAYS, AND SEND DATA AND PHOTOS BACK TO EARTH.

MARTIAN MISSION

At its closest, Mars is only about 50 million miles away from Earth! The Red Planet is, as far as we know, the only other planet in our solar system that might be habitable.

Aerospace engineer **DONNA SHIRLEY** (b.1941) knew we could get there. "I joined NASA in 1966 to work on Mars, I didn't actually get to do it until the *Pathfinder* mission came along in 1997. It took 30 years!" she says.

Donna was an integral part of the Mars *Pathfinder* mission to land a spacecraft on Mars, and then operate a rover around the planet's rocky surface. Her job was to build a small, light, and inexpensive rover that could travel within the *Pathfinder* probe.

Donna recalls, "This sounds easy, but in the late 1980s, rover designs were huge to accommodate computers that were the size of refrigerators." She and her team put together *Sojourner,* a micro-rover that is two feet long, one foot high, a lightweight 25 pounds, and packed with some serious exploration capabilities. The rover was able to check soil, measure dust, analyze the composition of rock, and take pictures.

The children of *Sojourner,* rovers called *Spirit* and *Opportunity,* are now motoring around Mars gathering even more scientific data.

TOP: 360-DEGREE PANORAMA OF THE MARS *PATHFINDER* LANDING SITE.

ABOVE: ARTIST'S CONCEPT OF THE ROVER ON MARS.

THAT'S THE SPIRIT!

Scientists and engineers had an outstanding opportunity to explore Mars as it moved close to the Earth in 2003. This time, they'd do it with something much bigger than the 25-pound *Sojourner* rover used in 1997.

Aerospace engineer **JENNIFER TROSPER** (b.1968) and her team of engineers and scientists built two rovers that could survive a trip through space and then maneuver on the Red Planet like robotic geologists.

Each rover weighs about 384 pounds, and is a little over five feet tall and a little less than five feet long. With communication systems, the rovers include all the tools needed to take pictures and gather information. Each rover was encased in a spacecraft that carried it to the Martian surface.

Spirit landed on Mars on June 10, 2003, and Jennifer became the *Spirit* mission manager. She says, "My job is to keep *Spirit* safe while trying as best we can to travel to areas that the scientists are most interested in. It is so exciting! Every time I hear and see the information sent from *Spirit* as it rolls along three million miles away, it reminds me that I've had a part in something that makes a difference."

So far, both *Spirit* and *Opportunity* (now operating on the other side of Mars) have provided extraordinary information about this seemingly barren landscape.

WHILE WATCHING FROM A CONTROL ROOM IN PASADENA, CALIFORNIA, JENNIFER TROSPER AND HER TEAM RIDE ALONG WITH THE *SPIRIT* ROVER AS IT BEGINS ITS JOURNEY AROUND MARS THREE MILLION MILES AWAY.

"Some folks say it's challenging for a female to be in the engineering field. But I guarantee you that if you focus on your competence and becoming the best possible engineer you can be, people will be lined up wanting to hire you, promote you, and be on your team."

—JENNIFER TROSPER

In Defense

Conflict is an inevitable part of civilization. Throughout history, nations have faced issues with strongly opposing views, a wide range of motivations, and varying ideals. The last 100 years have seen some extreme examples: from Adolph Hitler's desire to create a new world order to the recent deadly attacks by terrorist groups. These conflicts may bring an increased threat of attack—and even war.

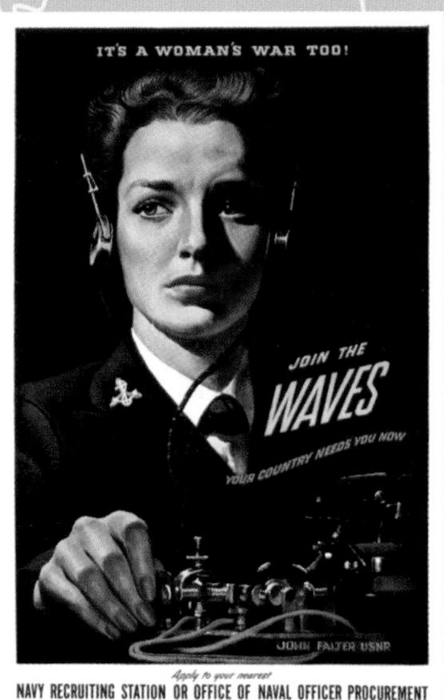

In wars, women and children have traditionally been the victims, suffering as widows or orphans, displaced from their homes, or imprisoned. And yet, there are many examples of women who have joined the fight for freedom.

Left: Navy warships head out to sea. Right: RADAR helps military land- and sea-based forces stay safe.

On the battlefield, women nurses and non-combat personnel have long been acknowledged for providing fighting soldiers with the best possible defense. In 1993, the United States opened the door for women to join men in battle.

On the home front, women—and women engineers—have moved into factories, laboratories, and research centers to develop new tools and technologies to keep their loved ones safe.

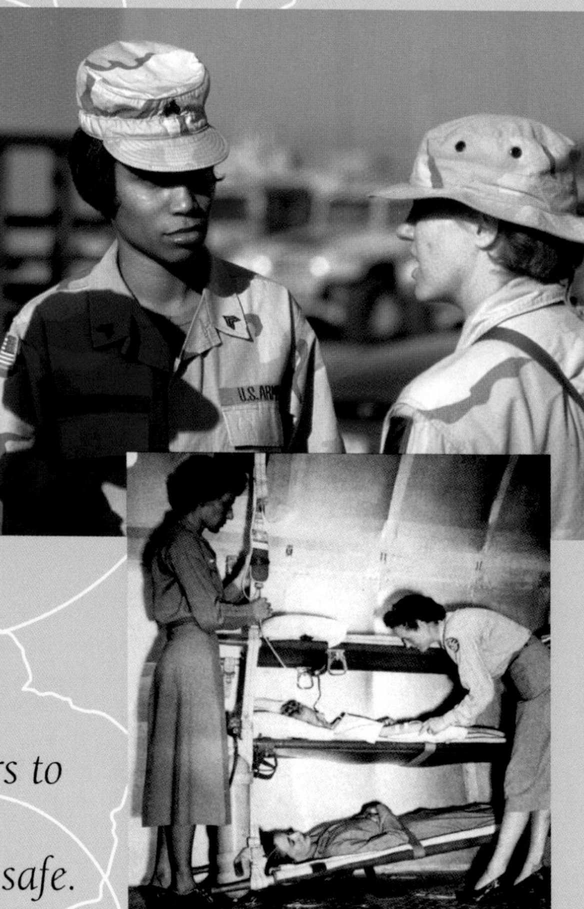

Women were allowed to officially join the military in WWII as nurses and support staff.

GREAT WORLD WARS

World War I (1914–1918) is called the "Great War" because never before had so many soldiers from so many countries been called to fight. Over nine million soldiers died in this four-year battle. Two decades later, nearly 20 million soldiers and 49 million civilians died during World War II (1937–1945).

During both wars, women were called on to serve in the military and work at jobs on the home front. Women like Alice Goff, Hazel Irene Quick, Lydia Weld, and Elsa Gardner took advantage of their skills in math and science to embark on new careers in engineering. By the end of World Wars I and II, a new image emerged of women able to "think" like engineers and provide innovative ideas to change our world—for the better.

"[Women] are in the shipyards, lumber mills, steel mills, foundries. They are welders, electricians, mechanics, and even boilermakers. They operate streetcars, buses, cranes, and tractors. Women engineers are working in the drafting rooms and women physicists and chemists in the great industrial laboratories."

NEWSWEEK MAGAZINE,
AUGUST 1943

LEFT: WOMAN WORKING ON AN AIRPLANE MOTOR AT NORTH AMERICAN AVIATION, INC., JUNE 1942.

RIGHT: THE CULTURAL ICON, ROSIE THE RIVETER, REPRESENTED THE MORE THAN SIX MILLION WOMEN THAT WORKED IN MANUFACTURING PLANTS DURING WORLD WAR II. THE WOMAN SHOWN IS RIVETING AN A-20 BOMBER ON THE ASSEMBLY LINE.

Prime Process

No matter what the product, chemical engineer **MARGARET HUTCHINSON ROUSSEAU** (1910–2000) found a better way to manufacture it. During World War II, she oversaw work in factories making penicillin and synthetic rubber, a key material needed for the nation's defense.

After the war, Margaret designed new equipment and improved methods for refining oil and producing chemicals such as ethylene glycol (anti-freeze) and glacial acetic acid, a nearly pure form of an acid used to make rubber, plastics, and pharmaceuticals.

Margaret Rousseau was more familiar with engineering as a career than most other women of her time. Her father was a petroleum engineer in oil-rich Texas.

Production of the B-17 Flying Fortress was possible because Dorothy Quiggle made 100-octane fuel that it could burn. The B-17 was the first mass-produced, four-engine heavy bomber built during World War II. It carried well over 2,500 gallons of fuel and had a range of 3,400 miles.

No More Knocks

For cargo and bomber pilots in World War II, better fuel meant more power and improved engine efficiency—so that the pilots could go to battle and come back quickly. Better fuel became the difference between winning and losing the war!

Chemical engineer **DOROTHY QUIGGLE'S** (1903–1993) specialty was fractionating hydrocarbons: breaking down crude oil into gasoline and other products. This expertise helped guide her development of 100-octane aviation gasoline. High-octane fuel allows a powerful piston engine to burn its fuel efficiently, a quality called "anti-knock" because the engine does not misfire, or "knock."

This new improved gasoline allowed the U.S. and Britain to build more powerful engines, and therefore bigger planes. The 100-octane gasoline and similar advances helped increase airplane top speeds by 15 percent and allowed planes to fly longer at higher altitudes. As a result of Dorothy's work, the U.S.'s aircraft engines performed better than similarly sized engines in the German Luftwaffe's air force, giving the Allies the edge.

Peace through Strength

Fresh from victory after World War II, the United States faced an unexpected problem: too many people for too few jobs. The thousands of men returning from war were heroes. They needed jobs. They needed opportunities for education and employment.

Women were encouraged to return to the home. While many women did indeed choose this route, there were others who either needed to work or simply liked the challenge their jobs provided. In continuing to work, they helped support our nation through a time of fear and uncertainty: the Cold War.

Engineers during this time produced some truly amazing and groundbreaking advances in technology, each designed to protect and deter possible attacks by another world power..

Launch of an Atlas missile from Cape Canaveral, Florida.

What Was the Cold War?

After the end of World War II, allied leaders from the United States, Great Britain, and the Soviet Union met to discuss the future of war-torn Europe. Unfortunately, the differing ideologies of these former allies created an even greater conflict.

Instead of cooperation, these meetings reflected a growing tension and distrust between leaders of the United States and the communist-run Soviet Union. These differences of opinion created a very different "battleground," often called the Cold War.

The two superpowers never fought an actual battle with each other. Rather, it was a battle of perceived strength. The subsequent 40-plus years was one of the most prolific periods of weapon and technology development in history.

Joseph Stalin, Franklin D. Roosevelt, and Winston Churchill meet to discuss post-war Europe at Yalta in 1945.

RESHAPING ROCKETS . . . AND BEYOND

Native American **MARY ROSS** (b.1908) joined Lockheed during the early days of World War II as a mathematician. She recalls, "They said they could make an engineer out of me if I'd work for them."

She was so skilled at her job, she was one of only 40 elite engineers to start Lockheed Missiles & Space Company. Mary says, "Over 30 years, I worked on everything from airplanes to rockets to interplanetary flybys. You can't get much more diverse than that!"

After she retired, Mary learned more about her Native American heritage. Her great-great grandfather was Chief John Ross, the man who helped lead the Cherokee Nation through the Trail of Tears. Mary joined the American Indian Science & Engineering Society and became a role model and mentor for Native Americans.

MARY ROSS SAYS HER WORK ON PLANETARY SPACE SYSTEMS WAS ONE OF THE MOST EXCITING AREAS OF RESEARCH. "OUR CHIEF ENGINEER WOULD COME BY AT THE END OF EACH DAY AND ASK US 'WHAT DID YOU DISCOVER TODAY?' THERE SEEMED TO BE UNLIMITED QUESTIONS—AND I HAD SUCH FUN DIGGING OUT NEW INFORMATION AND THEN PUTTING IT ALL TOGETHER TO FIND ANSWERS."

EARLY WARNING SAVES LIVES

With weapons in the Soviet Union becoming more sophisticated during the Cold War, the U.S. stepped up its efforts to develop air defense systems to protect its citizens from long-range attack. The Semi-Automatic Ground Environment (SAGE) system was one of the first designed to find, track, and then stop weapons-carrying aircraft.

SAGE relied on the most advanced radar, communication networks, computers, and computer programming technology available at the time. Radars located around the country would send bomber locations in over telephone lines to SAGE headquarters. There, the data were recorded and automatically analyzed to assess the appropriate response.

AILEEN CAVANAGH (1929–1992) made sure this complex radar system worked. She was closely involved in the engineering and shakedown of every element in the radar network. The SAGE system was fully operational in 1979, though concern had shifted from bomber threats to missile attacks. Thus, early warning systems such as SAGE were soon replaced by more advanced missile detection systems. Aileen's work on SAGE technology provided the foundation for an advanced air-traffic control system that was quickly adopted by the Federal Aviation Administration.

Aileen went on to help develop engineering studies for the Airborne Warning and Control System (AWACS) aircraft, the premier air battle command and control aircraft still used to support our military around the world today.

IN A 1962 AUTOBIOGRAPHICAL ESSAY, AILEEN CAVANAGH WROTE OF HERSELF: "INDEPENDENT OF MIND, DETERMINED OF SPIRIT, DEVOID OF FEAR, WITH A RESERVED MANNER BORN OF SHYNESS, AND STILL TORN BY CONFLICTING INTEREST BETWEEN THE FIELDS OF SOCIOLOGY AND TECHNOLOGY. SHE REBELS AGAINST SOCIAL CONFORMISM FOR THE SAKE OF UNIFORMITY OF THOUGHT; WAGES A CONSTANT BATTLE AGAINST FORCED TECHNICAL SPECIALIZATION FOR THE SAKE OF SELF-IDENTIFICATION AND EASE OF INDUSTRIAL OCCUPATIONAL CLASSIFICATION; AND BELIEVES IN THE LIBERAL EDUCATION AS AN ESSENTIAL INSTRUMENT OF SOCIAL DEVELOPMENT."

AT HOME ON THE RANGE

Fresh out of college with a degree in aeronautical engineering, **ARMINTA HARNESS** (b.1928) was told, "As long as a returning soldier needs a job, we're not about to hire a woman." Arminta looked to the Air Force for her engineering career. "It just seemed like a place where I might get a chance." She was right.

Over the next 24 years, she helped develop some of the nation's most significant weapons and space programs. One of her first projects was the B-29 Superfortress, one of the largest most advanced bombers of World War II. Later in her career, she helped design intelligence-gathering equipment for the U-2, the military's high-altitude surveillance and reconnaissance aircraft.

Arminta went on to manage the $2 billion Space and Missile Systems Organization. At the time, she's quoted as saying, "[This] may seem like a strange job for an engineer, but that's one thing I enjoy about the work. There's a good mixture of financial and technical people."

Arminta is now retired as a Lieutenant Colonel in the Air Force.

"I became an aeronautical engineer because of some advice I read in Amelia's [Earhart] books, plus my love of both math and flying. In fact, I fell in love with flying at age three when I saw my first airplane and its 'barn storming' pilot at a field near Amarillo, Texas. Years later I earned my own pilot's license—a few years before I could drive."

—ARMINTA HARNESS

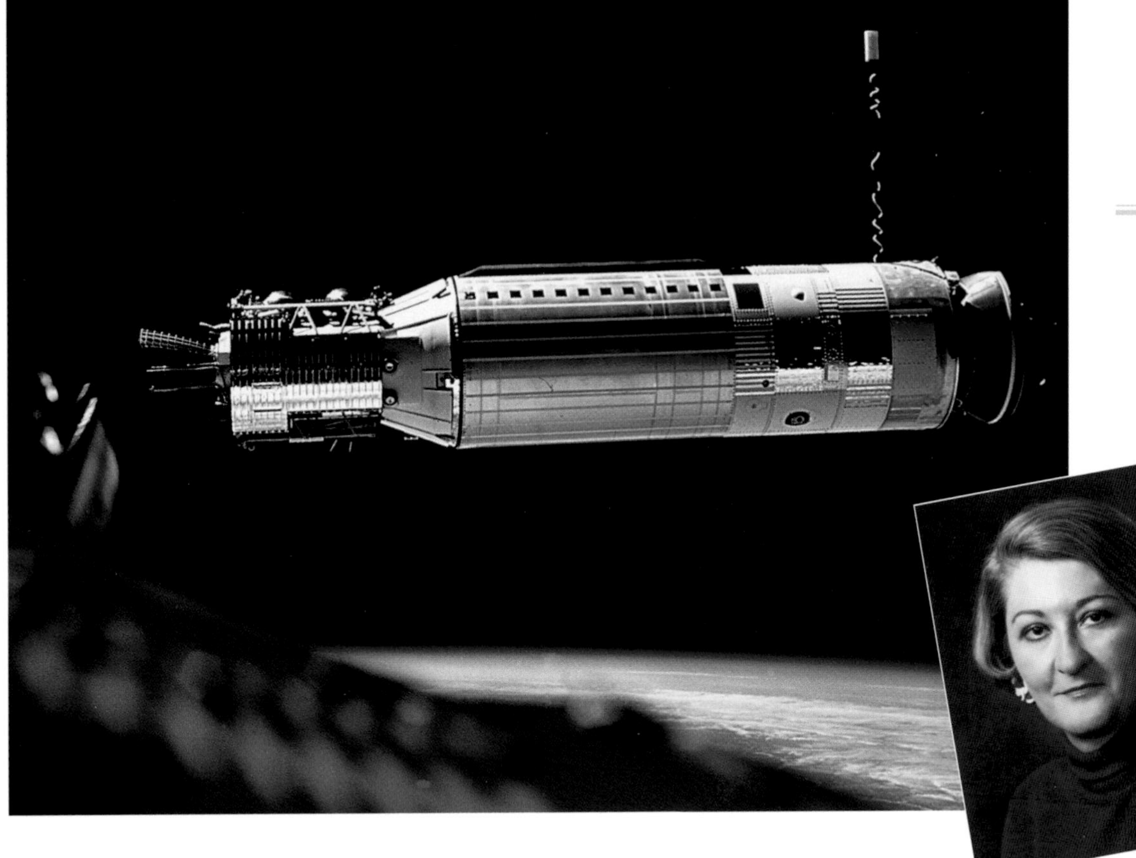

DURING HER MILITARY CAREER, AS PART OF THE GEMINI MANNED SPACE PROGRAM IN THE 1960S, ARMINTA HARNESS HELPED DRIVE THE FIRST EFFORTS TO PUT PEOPLE IN SPACE. SHE AND HER COLLEAGUES BUILT AN UNMANNED SPACE STATION THAT COULD SUPPORT ASTRONAUTS WHILE THEY MOVED ABOUT SPACE, SETTING THE STAGE FOR LANDING ON THE MOON AND SPACE EXPLORATION.

WATCHING OVER WAR FIGHTERS

As Saigon, Vietnam, was immersed in smoke and fire, 15-year-old **ANH DUONG** (b.1960) and her family escaped with the clothes on their backs, leaping aboard one of the last helicopters to leave war-torn South Vietnam before it was overrun by the North Vietnamese army.

Anh and her family were lucky. Her brother, a South Vietnamese Air Force officer, was the pilot. The helicopter flew to a nearby naval ship where Anh and the rest of the refugees were whisked away to a camp in the Philippines, then to another one in Pennsylvania.

Supported by the generous hearts and helping hands of a local church, Anh's family had a new home and a new start. While the adults in her family looked for jobs, Anh entered the local high school knowing only 50 words of English. Her sophomore school year was more than half over and she was way behind. Two years later, she graduated in the top three percent of her class—receiving an "A" in her honors English class—and went on to college.

While her English had improved dramatically, she now also discovered a knack for math and science. When she graduated *cum laude* in chemical engineering, Anh wanted to use her skills, and her heartfelt gratitude, to serve the soldiers—"war fighters," she calls them—whose sacrifice earned her a new life in the U.S.

"The war fighters kept me and my family safe and free then, and they continue to keep us safe and free here. I want to repay my immense debt to them," Anh says. As an engineer with the Naval Surface Warfare Center, Anh leads the nation's top developers of high explosives—the materials that go into bombs, missiles, and torpedoes and make them blow up.

THE VIETNAM WAR PITTED THE COMMUNIST NORTH VIETNAMESE (BACKED BY CHINA AND THE SOVIET UNION) AGAINST THE DEMOCRATIC SOUTH VIETNAMESE (BACKED BY THE UNITED STATES). THE U.S. WITHDREW ITS FORCES IN 1973, BUT THE WAR CONTINUED. UNDER HEAVY FIRE FROM COMMUNIST GUNNERS, RESCUE HELICOPTERS TRIED TO EVACUATE PEOPLE TO SAFETY.

ANH DUONG AND HER TEAM DEVELOPED THE THERMOBARIC BOMB IN JUST 67 DAYS TO SUPPORT TROOPS IN AFGHANISTAN. IT IS USED BY TROOPS TO INFILTRATE CAVES AND TUNNELS.

Nationally and internationally recognized as an explosives expert, Anh was entrusted with a special mission in the wake of the September 11, 2001, terrorist attack: the creation of the first U.S. thermobaric bomb, a bomb capable of diving deep into caves and tunnels to destroy enemies' facilities inside.

"This type of weapon is designed to spare our soldiers in Afghanistan the dangerous task of taking on the terrorist defenses on foot in unfamiliar territory," Anh says. Anh's team of nearly 100 scientists, engineers, and technicians completed their mission in a record time of 67 days, earning numerous Navy awards and citations.

Today, Anh serves as the director for science and technology at the Naval Surface Warfare Center, Indian Head Division. "Part of my job is to anticipate what our war fighters need to fight future wars. While it's a tremendous challenge, I'm honored to have opportunities to serve the men and women who gave me and my family a second chance."

Protection Today

Not long ago, the threat of large-scale invasion in the U.S. seemed an unimaginable possibility to most of us. That changed in 2001 with the attacks on New York's World Trade Center and Washington D.C.'s Pentagon. Now safety, security, and homeland defense are an integral part of our everyday language. Terrorism is a real and constant threat.

Homeland defense—securing people and communities from terrorist attacks—relies heavily on technologies like remote control robots, miniaturized detection devices, and satellite-based defense systems rather than conventional weapons such as guns and missiles.

The military is also shifting to manage this new era of national security more effectively while still providing global assistance to those in need. Women engineers have been called on to drive that transition, to become defenders of the peace and of their homeland.

"For it isn't enough to talk about peace.
One must believe in it.
And it isn't enough to believe in it.
One must work at it."

Eleanor Roosevelt
First Lady, Human Rights Advocate, Author, and Humanitarian (1884–1962)

The Predator is an unmanned, remotely piloted aircraft used to conduct surveillance and reconnaissance missions by the U.S. Air Force in areas that are unsafe for our military forces. Powered by a turbo-charged engine, this 1,130-pound aircraft holds cameras, satellite links, radios, and even de-icing features. Data gathered by the Predator is instantly available to U.S. forces with the help of large military satellite dishes located around the globe.

Background: A sattelite orbits earth.

MILITARY LEADER

For decades, the United States Air Force has been about airplanes—designing, building, and flying some of the most advanced aircraft in the world. Now, this elite force is setting its sights on space, relying on communication satellites, space-based alert systems, and satellites equipped with technology that can pinpoint the location of any place on Earth in just minutes.

Coming up with a new plan to meet the needs of the 21st century began in the 1990s, and aeronautical engineer **SHEILA WIDNALL** (b.1938), then Secretary of the Air Force, helped shape the long-range vision.

Sheila says, "Being an engineer was enormously important in so many ways in my role. Because I was able to understand the basic functionality of air and spacecraft and weapons systems across an enormous swath of technologies and systems—even though I wasn't an expert in all of them—I was better able to communicate my ideas.

SHEILA WIDNALL was the first woman to head a military service (1993 –1997).

"In my position as the head of the Air Force, I made sure that our forces were ready to go anywhere to defend America's interests—often non-military related. Many people don't realize that about half of the Air Force's capabilities are used on a routine basis for humanitarian missions," Sheila says.

Sheila points to two events in particular. The first was flooding in Mozambique, Africa, in 2000 that displaced nearly one million people.

"Relief organizations couldn't get anywhere near the country to help because of the flooding. We arrived in helicopters with water purification systems and other needed supplies within just days. It was a major relief effort."

The second was during the uprisings in Somalia, Africa, in 1994. It was up to the military to move U.S. citizens and civilians from one city to another to get them out of danger.

"Our role as a major superpower is to bring aid and defense to many parts of the world. To do this, we need to make sure that we have the technology to provide this support, and also to make sure that the families left back home are taken care of. When people make sacrifices to serve their country, they shouldn't have to worry about their families having safe and good-quality living conditions, good schools, and all necessary medical care."

SHEILA WIDNALL (STANDING, AT RIGHT) SAYS THE HANDS-ON LESSONS LEARNED FROM HER EARLY COLLEGE DAYS STUDYING ENGINEERING SET THE FOUNDATION FOR HER FUTURE—AND FOR A LOT OF UNUSUAL EXPERIENCES. AS SECRETARY OF THE AIR FORCE, SHE FLEW IN THE MOST ADVANCED AIRPLANES IN THE WORLD, LANDED ON AN AIRCRAFT CARRIER, AND WORE A SPACESUIT TO FLY ON THE U2 AT 65,000 FEET WHERE THE AIR IS BLACK BUT YOU CAN SEE THE CURVATURE OF THE EARTH.

A Humanitarian Tradition

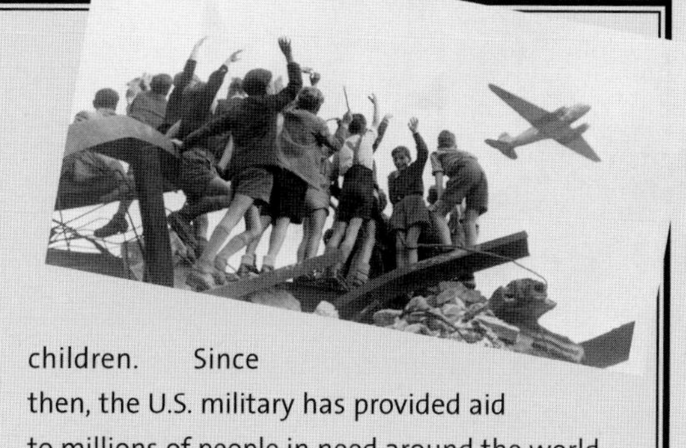

The 1948–1949 Berlin Airlift defined the Air Force legacy of humanitarian relief during natural disaster or political unrest. For 462 days, Air Force pilots dropped over two million tons of supplies into communist-blockaded Western Berlin. Cargo ranged from coal for heating households to packets of candy for children. Since then, the U.S. military has provided aid to millions of people in need around the world.

ANTICIPATING THREATS

Aerospace engineer **SUZANNE JENNICHES** (b.1948) has made huge contributions to several extraordinarily high-profile projects directly related to national security.

Her engineering career began in 1974 and she progressed through a number of test and manufacturing assignments. Her first engineering management contribution was to the B-1B Lancer strategic bomber built in the 1980s. The B-1B was built to fly great distances, very quickly, undetected. A bomber is not often thought of as a defensive weapon, but in this case the B-1B's capabilities were so powerful that it acted as a deterrent. The B-1B helped stop the Soviet Union from making further advances at the end of the Cold War.

Suzanne built an electronically scanned antenna for the B-1B offensive radar that, when working with other elements of the radar, provides critical information such as a high-resolution land map, location information, weather data, and much more. Suzanne recalls, "This had never been done before but was really critical to the whole mission. It's the only way for the plane to automatically maneuver while flying under enemy radar."

Today, Suzanne is again at the center of several critical projects—this time in homeland defense and robotics. "This work is so crucial to our society and the American people, particularly since the terrorist attacks on September 11, 2001. It's a tremendous challenge because we're trying to predict the next anticipated threat—and these threats change continually."

SUZANNE JENNICHES IS SHOWN HERE WITH BIO-DETECTION MAIL SORTERS DEVELOPED FOR THE U.S. POSTAL SERVICE. INSET: THE ANTHRAX BACTERIUM SPORE IS CONSIDERED A BIOLOGICAL WEAPON.

RIGHT: SUZANNE ALSO MANAGES A GROUP THAT BUILDS MOBILE SATELLITE COMMUNICATION DISHES. THESE DISHES ARE EASY TO SET UP ANYWHERE IN THE WORLD, AND ARE ABLE TO INSTANTLY RECEIVE DATA FROM SATELLITES THAT CIRCLE THE EARTH.

Soon after the World Trade Center towers fell, U.S. citizens faced another, equally scary threat: anthrax. Someone put powder containing the deadly bacteria in envelopes and mailed it through the U.S. Postal Service. Anthrax is so lethal that just a few spores containing DNA can kill. During the anthrax attacks in 2001, 11 people inhaled anthrax and only six survived. An additional 17 people developed skin sores from anthrax exposure.

Suzanne is especially proud of her team's work during the anthrax attacks. Just six months after the first attack, they rolled out a powerful, affordable detection system that could be installed quickly and easily at any postal facility.

MIGHTY MINI-TECH

Sandia National Laboratories helps the U.S. "secure a peaceful and free world through technology." Since 1974, mechanical engineer **JOAN WOODARD** (b.1952) has worked at Sandia where she's had the chance to really make a difference. "Not only in national security," says Joan, "but also in improving the quality of life for the general public. For instance, I've been able to drive technology that can improve health care with tools that are less invasive."

Today, she's hard at work on miniaturized technologies that pack a lot of power. Airports and border patrol agents already use some of the tools Joan has helped develop. For example, the lab team has developed an explosives detection machine that senses the presence of bombs. It works much like a metal detector. People walk through the portal, which puffs out a blast of air, and then collects a sample. From this sample, the system can tell whether the person is carrying explosives.

"Just one idea rolled out to airport terminals around the world has the ability to protect —and sometimes save—the lives of millions of people. That's a good day's work!"

LEFT: JOAN WOODARD SPEAKS AT A "STATE OF THE LABS" PRESENTATION WHERE SHE SUMMARIZES THE TECHNOLOGIES OF TOMORROW THAT ARE CURRENTLY UNDER DEVELOPMENT AT SANDIA NATIONAL LABORATORIES.

ABOVE: IN HER SPARE TIME, JOAN ENJOYS WORKING IN THE COMMUNITY. HERE, SHE GETS TO SPEND THE DAY WITH YOUNG STUDENTS AT THE SANDIA BASE ELEMENTARY SCHOOL DURING "READ ACROSS AMERICA DAY."

CHERYL BERGMAN SAYS THAT WHEN LOOKING FOR A JOB, SHE MADE SURE TO FIND A COMPANY AND MANAGERS THAT WERE WILLING TO WORK WITH HER AND OTHER EMPLOYEES TO MEET THE NEEDS OF CAREER AND FAMILY. TODAY, CHERYL SPENDS PART OF HER TIME WORKING IN FLIGHT SIMULATORS TO HELP PILOTS FLY MORE SAFELY, AND THE OTHER PART WITH HER CHILDREN ON THE SOCCER FIELD.

SUPERSONIC SIMULATIONS

The next-generation stealthy, supersonic aircraft, the Joint Strike Fighter (JSF), will have the mark of many women. Electrical engineer **CHERYL BERGMAN** (b.1966), for example, is in charge of the JSF Virtual Cockpit Simulation Team. "My job is to make it easier for pilots to use complex technologies," says Cheryl.

Cheryl builds the JSF's pilot-vehicle interface—the bridge between the airplane's flight and weapons technology and the fighter pilot. "We basically build a flight simulator and then bring the pilots in to use it. As the pilots work through various battle scenarios, we're able to see how they interact with the technology. We're then able to make improvements. We want the interface to work as naturally as possible, so that when the pilots go into battle, the airplane can help them accomplish their mission and stay safe."

Through simulation, Cheryl improves the pilot-vehicle interface design with easier-to-read displays, better equipment, and even better ways for the pilots to control the aircraft during dangerous life-and-death situations. "Engineering is so much more than design—it's a window of opportunity."

Robots to the Rescue

Robotic soldiers? Yes! Robots are an increasingly important addition to a soldier's arsenal. Many of them are the creation of mechanical engineer **HELEN GREINER** (b.1967), chairman and co-founder of iRobot.

"It's extremely gratifying to know that our robots protect and save lives," Helen says. "Since I was 11 years old, I knew I was going to study engineering so that I could build robots. People told me I should stick to something more practical, but I really believed I could make a difference."

Her military robot, called "iRobot PackBot," can walk, respond to orders, climb up stairs, fall down a cliff and get back up, and even operate in severe weather conditions. The PackBot Tactical Mobile Robots wear different "backpacks," depending on the soldier's mission. Some backpacks contain cameras or radios; others contain sensors that can tell whether harmful chemicals are nearby. Some can even disconnect bombs.

Helen's company is also developing a new generation of unmanned ground vehicles for the Army. The FCS Small Unmanned Ground Vehicle will be a smaller, lighter successor to the combat-proven iRobot PackBot Tactical Mobile Robots.

Also underway is a line of commercial robots that do everything from vacuuming to scrubbing floors. Helen adds, "I'm doing what I love and at the same time making a difference in the world—what more could I ask from a career?"

Helen Greiner creates robots for both military and commercial applications. Above, she's shown with the popular iRobot Roomba Vacuuming Robot that automatically cleans your entire house, at the touch of a button. At right is a robot used for military missions.

NUCLEAR SAFETY

During the Cold War, several countries, including the U.S., the Soviet Union, and China, developed nuclear weapons as a military deterrent. Nuclear weapons are able to level entire metropolitan areas and leave behind extremely high radiation levels that are deadly to humans for years to come.

In the 1960s, there was broad fear that the availability of these weapons would trigger a global nuclear war. This fear prompted intense negotiations amongst world leaders. The negotiations resulted in nuclear non-proliferation treaties that restrict which countries own nuclear weapons and also restrict the use of nuclear weapons.

Mechanical engineer **CHRISMA JACKSON** (b.1973) makes sure that the U.S. nuclear bombs and warheads stay safe and secure. Chrisma helps upgrade the electrical and mechanical systems that protect these weapons. She applies emerging technological advances to develop new ways to protect them.

Chrisma says, "It's so important that we keep these weapons safe from our enemies. I apply my engineering knowledge every day to develop ways to keep the weapons safe and secure. I'm very proud of my job and enjoy doing work for the good of the country."

INSET: CHRISMA JACKSON VISITS WHITEMAN AIR FORCE BASE IN MISSOURI TO INSPECT A BOMBER'S LAUNCHER.

RIGHT: CHRISMA GETS READY TO CONNECT THE MAIN POWER CONNECTORS TO TEST THE SAFETY AND SECURITY FUNCTIONS OF THIS WARHEAD.

A PENTAGON POWERHOUSE

As an engineer and computer scientist, **ANITA JONES** (b.1942) has steered a submarine, landed in a jet fighter on an aircraft carrier, trekked to the South Pole, and jumped off a parachute training tower.

In 1994, she became the director of defense research and engineering for the Department of Defense (DOD). Anita recalls, "Our job was to develop and apply innovative science and technology to help the military. We delivered unpiloted planes, better-tasting food, and more intelligent computer software." Her job with the DOD has been completed but Anita regularly serves on scientific boards that advise the Pentagon and civilian agencies about problems in which technology may be part of the solution. Currently, she chairs a high-level committee to advise Congress on what to do about aging polar icebreaker ships.

Says Anita, "If you become an expert in some field of engineering, people will come to you. They open doors and ask you to work with interesting people while creating useful solutions to society's problems."

ONE OF THE MOST EXCITING PROJECTS FOR ANITA JONES AND HER SCIENCE AND TECHNOLOGY TEAM WAS THE IMPLEMENTATION OF UNMANNED AIR VEHICLES OR UAVS, SMALL AIRCRAFT TYPICALLY USED TO GATHER DATA IN WAR ZONES. "MILITARY COMMANDERS TESTED A COUPLE OF EARLY PROTOTYPES. THEY LIKED THEM SO MUCH, THEY WOULDN'T GIVE THEM BACK!"

Protecting the People: A Day in the Life of Amy Alving

The September 11, 2001, strikes on the World Trade Center and the Pentagon demonstrated the extreme vulnerability of U.S. infrastructure to terrorist attack. Finding ways to reduce this vulnerability is the focus of organizations like Defense Advanced Research Projects Agency (DARPA). And DARPA is Amy Alving's (b.1962) home away from home.

Since her college days, Amy wanted to influence the direction of U.S. technology policy. Through education and experience, she has learned to work effectively inside Washington, D.C.'s political structure, while helping to guide tomorrow's defensive technology such as anthrax detection systems and unjammable communication networks.

Amy begins each day with one goal: Find ways to use advanced technology to protect those in danger. Here's how she puts ideas into action.

"I often begin my day briefing colleagues or talking with the Director of DARPA. For example, I might discuss progress on the Immune Buildings Program. My team helped develop early-warning technology that detects the presence of harmful chemical or biological substances that could enter a building. The technology then reacts intelligently to protect people. That can save lives!"

In her free time, Amy loves to travel around the world to see historical and culturally fascinating places, such as beautiful Avignon, France (far left). Her job takes her around the world as well. She enjoys meeting with friends from various countries for lunch, dinner, or even some recreation. Here, on a day off in Singapore, she joined some of her colleagues on the golf course.

Much of Amy's job is spent speaking to other military support groups, such as the Air Force Space Command. Sometimes it's in an F-16 fighter jet (far left). "The technical scope of this job is astonishing, but there are so many aspects beyond just this. I must anticipate emerging threats, interact with many branches of the military, and coordinate with government organizations. All of these activities exercise different skills, so this job is never boring!"

"I'm an intermediary between the engineers who make defense technology and the politicians who make decisions about how the public and the government want to spend money. During my conversations with congressional representatives, I have to convince them that projects are worthwhile to fund."

In 2004, the Service to America selected Amy as one of four finalists for a Science and Environment Medal in recognition of her efforts to develop special sensors that detect and block chemical or biological agents in a building's ventilation system.

"Part of my job is to think about how our technology can make a difference. this could mean evaluating U.S. satellite communication network security or developing defense systems to prevent a space-based Pearl Harbor."

> *"Every day is fast-paced and packed with ideas, solutions, and possibilities. I can't wait to get to work!"*
>
> —Amy Alving

GATEWAY TO THE WORLD

The 21st century promises many challenges and worlds of opportunity waiting to be explored, worlds as tiny as a single human cell to as large as the universe. Worlds within ourselves. Worlds at our doorstep. And worlds beyond.

In the past, boys and men have typically been the ones to explore these worlds. But women engineers have broken through those barriers—which makes it easier for women today to follow their own paths.

Teams of engineers from all walks of life are needed to work together to better our lives. These teams need infusions of creativity, new approaches, and different ways of looking at

the world. In short, these teams need women engineers!

A career in engineering can be enormously creative—and can serve as the gateway to all sorts of amazing careers that may not typically be considered "engineering." Women engineers have successfully become leaders in government, business, medicine, and law.

A FAMILY HARVESTING TOMATOES IN ZIMBABWE.

They have applied their engineering skills to tackle some of the world's most challenging problems: pollution, hunger, and human rights. In doing so, they have made a difference. Their lives stand as shining examples of how individual women, empowered with the proper skills, can help improve the dignity and prosperity of all the world's peoples.

Trailblazers, Pathfinders, Scouts

Engineering is a passionate profession full of brave and determined women who loved engineering so much that they fought for the right to practice it. Early, pioneering women engineers endured many minor inconveniences, such as not having a women's restroom in their workplace. They also endured larger issues, such as disrespect, hostility, and flat-out rejection.

Their strength and determination prevailed in the workplace and in the engineering profession. Today, women engineers are fully embraced by the profession. Knowing what may lie ahead for young women entering the field of engineering, many trailblazing women engineers have dedicated their life's work to making the path easier to follow.

"If I have seen further,
it is by standing on
the shoulders of giants."

SIR ISAAC NEWTON
MATHEMATICIAN AND PHYSICIST (1643–1727)

For Future Generations

With passage of the 19th amendment to the U.S. Constitution on May 19, 1920, American women finally won the right to vote. Sixty years later, structural engineer **RUTH V. GORDON** (b.1926) chained herself to the Pacific Stock Exchange to protest gender discrimination and to demonstrate that the Equal Rights Amendment was an economic issue. "The suffragettes did it for me," she explains. "I figured I could do it for my children." (Read more about Ruth's engineering career on page 45.)

ENGINEER AND FIREBRAND

Civil engineer **NORA STANTON BLATCH** (1883–1971) first campaigned for suffrage—a women's right to vote—at Cornell University, where she chose civil engineering as her major because it was the most male-dominated field she could find. She graduated in 1905.

That same year, Nora was the first woman admitted (although with "junior" status) to the American Society of Civil Engineers. In 1916, despite Nora's success as an architect and civil engineer, the ASCE terminated her membership because she passed the age limit for junior status.

Nora brought her case to the Supreme Court, claiming that ASCE rejected her application for full membership not because she was professionally incapable, but because she was a woman. She lost the case. It would be 11 more years before a woman, Elsie Eaves, gained full membership in ASCE, and another 76 years before a woman, Patricia Galloway, became a president of ASCE.

MOTHER'S AN ENGINEER

"A very nice, well-to-do lady paid for my father to go to college," explains chemical engineer **MARYLY VAN LEER PECK** (b.1930). Maryly's father, Blake Ragsdale Van Leer, went on to become dean of engineering at the University of Florida and North Carolina State, and then president of the Georgia Institute of Technology. "I've spent my life trying to give others the same opportunity."

One of Maryly's lasting legacies as the first woman president of the Polk Community College was to establish a number of endowed scholarships. Maryly is largely responsible for organizing the framework for the Society of Women Engineers' student sections, and helped to organize many of the early student sections. She is an active supporter of Girls Incorporated, which teaches young women to be the best that they can be, whatever that may be.

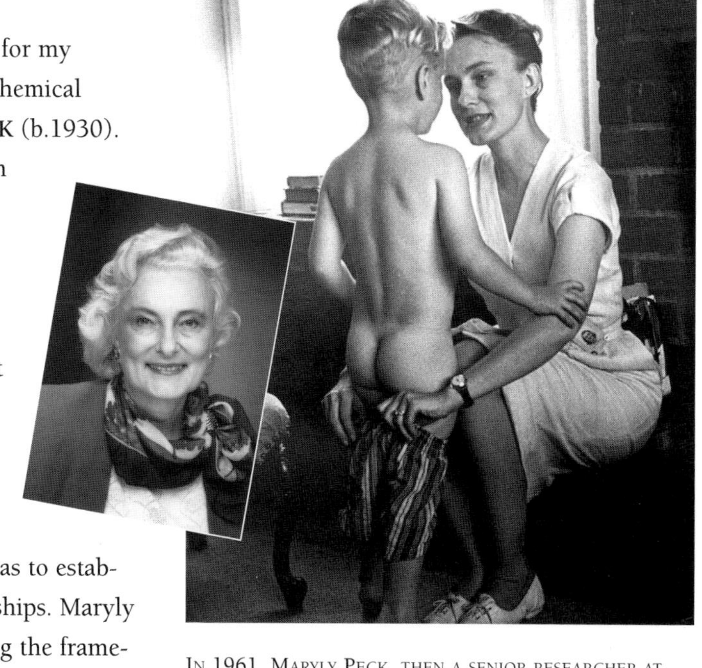

IN 1961, MARYLY PECK, THEN A SENIOR RESEARCHER AT ROCKETDYNE WORKING ON HYBRID FUEL COMBUSTION, WAS FEATURED IN *LIFE* MAGAZINE—ALONG WITH 100 OF THE MOST IMPORTANT YOUNG MEN AND WOMEN IN THE U.S.—AS ONE OF THE TAKEOVER GENERATION. "MOTHER'S AN ENGINEER," READ THE HEADLINE. THREE OF HER FOUR CHILDREN, HER NIECE, AND THREE NEPHEWS ARE NOW ENGINEERS.

IMAGINE BEING THE DAUGHTER OF HARRIOT STANTON BLATCH (1856–1940) (RIGHT) AND THE GRAND-DAUGHTER OF ELIZABETH CADY STANTON (1815–1902) (CENTER), BOTH FAMOUS LEADERS IN THE EARLY WOMEN'S RIGHTS MOVEMENT IN THE U.S. NORA STANTON BLATCH'S (LEFT) FATE AS A FEMINIST WAS SEALED!

Engineering Needs Women

A good sense of humor helped electrical engineer **ELEANOR BAUM** (b.1940) put up with others who didn't take her career seriously, that, and challenging the status quo of the engineering profession. "An individual does not need to be a white male to succeed in engineering," she says.

After a number of unsatisfactory industry jobs, Eleanor went back to school for her Ph.D. in electrical engineering. She eventually became the first female dean of engineering in the United States. She is now the dean of engineering at The Cooper Union in New York, making sure that her university provides an environment that is welcoming and nurturing to women's success. "Any creative endeavor that excludes a cross-section of the population is losing something. Engineering needs women!"

Eleanor Baum has worked tirelessly to encourage minorities and women to pursue an engineering education, and has influenced institutions around the world to welcome women.

"Not Fair!"

A lot has changed since the early days when women first started studying and practicing engineering. Twins **MARY LOCKETT HUTSON** (1884–1982) and **SOPHIE PALMER HUTSON** (1884–1983) were allowed to enroll at Texas A&M—60 years before the university opened its doors to women—because their father was a professor there. They studied civil engineering and completed all requirements in 1903.

But because they were women, they were issued "certificates of completion," not true diplomas: in short, second-rate acknowledgement for their first-rate education. Mary went on to design several water pumping facilities around New Orleans and in Mississippi and Texas. It wasn't until 2002 that the university conferred degrees on the twins—posthumously.

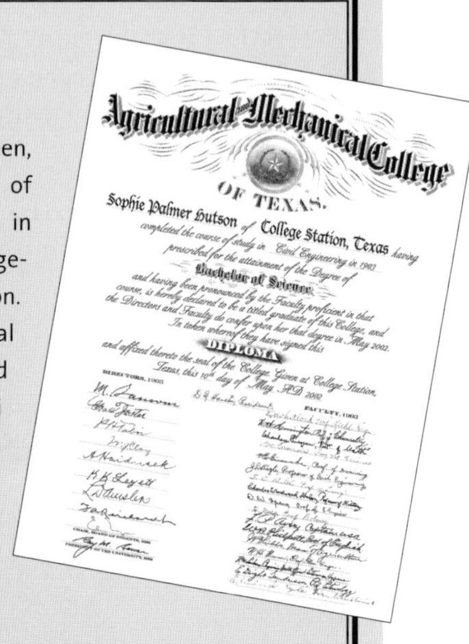

Agricultural and Mechanical College of Texas.

Sophie Palmer Hutson of College Station, Texas having completed the course of study in Civil Engineering in 1903 prescribed for the attainment of the Degree of Bachelor of Science and having been pronounced by the Faculty proficient in that course, is hereby declared to be a titled graduate of this College, and the Directors and Faculty do confer upon her that degree in May 2002. In token whereof they have signed this Diploma and affixed thereto the seal of the College. Given at College Station, Texas, this 10th day of May, A.D. 2002

Directors, 1903

Faculty, 1903

Chair, Board of Regents, 2002

President of the University, 2002

From Solo to Teamwork

In graduate school, electrical engineer **IRENE CARSWELL PEDEN** (b.1925) felt extremely isolated. It was a "long, lonely path," she says about being the first woman to obtain a Ph.D. in engineering from Stanford University, but she did it!

Although Irene thrived on the quantitative nature of engineering, she was quite people-oriented, which may explain her attraction to teaching. "Back then, engineers worked alone to solve problems. Today, we know that we need to work in teams. Engineering education now reflects this."

To this end, in addition to a distinguished career as a teacher and researcher of electromagnetics, wave propagation, geophysical remote sensing, and antennas, Irene has applied her engineering style to bettering the university environment for women—through supporting, enabling, and mentoring other women. "So much of my work was done in the old male-dominated "Lone Ranger" environment once typical of engineering. We're fortunate that engineering education has changed to accommodate all learning styles and needs, not just men's."

For five years, Irene Peden relied on data about the properties of deep ice collected by her male colleagues. There's no substitute for obtaining information first-hand, though. In 1970, Irene (right with Julia Vickers, an assistant from New Zealand), was the first woman principal investigator to conduct field studies in the Antarctic. To this day, Irene swears that they overcame sabotage to complete her research there.

Government for the Women

In his 1863 Gettysburg Address, Abraham Lincoln described the democracy of the U.S. as "government of the people, by the people, [and] for the people." At the time, "people" meant men. Women were not allowed to vote, and they were certainly not allowed to hold elected positions in the government.

In 1916, Montanan Jeannette Rankin was the first woman in the United States to serve in Congress. Women won the right to vote in 1920 with the passage of the 19th amendment to the U.S. Constitution.

If a government is an organization that has the power to make and enforce laws, and roughly half are women, why aren't women equally represented in that government? It's one of the tough social questions that women and men have been struggling with for decades. Could the know-how of engineers help make the government workplace more equitable for women? Certainly!

"Because man and woman are the complement of one another, we need woman's thought in national affairs to make a safe and stable government."

Elizabeth Cady Stanton
A Leader of the 19th Century American Women's Rights Movement (1815–1902)

Women cast their first votes for President in November 1920.

Champion of Equality

In 1961, president John F. Kennedy asked former first lady Eleanor Roosevelt to chair the first Presidential Commission on the Status of Women. Civil engineer **EVELYN "EVIE" BARSTOW HARRISON** (1910–2000) worked closely with Mrs. Roosevelt on the commission to investigate questions regarding women's equality in education, in the workplace, and under the law.

In 1963, Evie became the head of the newly formed Federal Women's Program, part of the Civil Service Commission, with a goal of making the federal government the nation's model employer of women. She advocated for full equality for women, and used her position in the Civil Service to further initiatives that supported the rights women enjoy today.

Evelyn Harrison (front, center) was the first woman to graduate from the University of Maryland with an engineering degree. Like many women in this book, Evelyn was the only woman in a class of men.

Modernizing Military Support

After spending most of her career developing weapons systems, mechanical engineer **MARY LACEY** (b.1955) took over the National Security Personnel System (NSPS) to overhaul the way Department of Defense civilian employees are hired, paid, and promoted.

The question Mary gets asked most often? "Why put an engineer in charge of the NSPS?"

She responds, "It's all about setting up a system that enables people to complete the work of the organization. Designing a personnel system is not much different than designing a complex weapons system or Navy ship. As an engineer, I know how to set up systems that work, and this job plays to my passion. I love people and I like change."

As military strategies change to meet new threats of the 21st century (for example, stepped-up national security and the global war on terrorism), the civilian workforce must be able to respond efficiently and quickly to meet military needs.

"A strong and flexible military is necessary to maintaining our way of life," Mary says. "As an engineer, I've been able to support that cause, whether through technology development or people management."

Mary Lacey is using her engineering skills on her most complex project yet: overhauling the Department of Defense's civil service personnel system.

Getting Down to Business

Since the beginning of humankind, enterprising people have bought, made, and sold goods and services—in short, created businesses. Engineers are no different. In fact, engineers' problem-solving skills are an ideal basis for success in the business world!

Running a business involves identifying a need (in business parlance, a "market"), envisioning a solution (a "product" or "service") to fill that need, and gathering and organizing the resources ("materials" and "personnel") needed to create the product or service. This is just what engineers do!

An Engineer on Wall Street

Mechanical engineer **KRISTIN STOEHR PEREIRA** (b.1965) compares business to machines. Each business process, whether finance or sales, has "switches" and "motors" and "gears" that make it work. If the motor doesn't have enough power or the gears don't mesh, the business process won't work smoothly or efficiently.

Kristin started her career with General Electric Aircraft Engines, moved to General Electric Plastics, and became one of General Electric's first "black belt" experts at Six Sigma, a highly disciplined process that helps businesses focus on developing and delivering near-perfect products and services.

Today, Kristin is using her engineering skills in the world of high finance. Her company, the Financial Guaranty Insurance Company (FGIC), provides insurance for many high-stakes Wall Street bonds and securities sales. As director of operations for FGIC, Kristin helps streamline the switches, motors, and gears that drive the business.

"Being able to think through a problem and develop 'if-then' scenarios makes it easy to understand all sorts of different business processes," says Kristin Pereira. Her office is in midtown Manhattan (New York City) overlooking Grand Central Station.

A Woman Before Her Time

Automaker Henry Ford credited **KATE GLEASON** (1865–1933), not her father who was the founder of the Gleason Machine Tools Company, for designing and perfecting a machine that produced beveled gears quickly and cheaply. Ford said that it was "the most remarkable machine work ever done by a woman," even though Kate had only a few months of formal mechanical engineering training at Cornell University.

As secretary-treasurer of the Rochester, New York, firm from 1890 to 1913, Kate's talented leadership helped the small, family-run company grow into a prosperous and nationally prominent producer of gear-cutting machinery. She then turned her business skills toward restoring another machine-tool company to financial solvency, became president of the National Bank of East Rochester, and then became a builder of affordable housing and a land developer.

Kate left an estate of $1.4 million at her death—worth around $18 million today! Much of the money was bequeathed to philanthropic causes, including the Rochester Institute of Technology—where today, roughly 500 students graduate from the Kate Gleason College of Engineering each year.

Above: One of Kate Gleason's businesses, the Concrest Community, built affordable houses for working people in East Rochester. People still live in the houses today.

Above: Kate Gleason (far right) started helping out in her father's machine shop at the tender age of 11 and became the bookkeeper by age 14.

Jet-Setting Problem-Solver: A Day in the Life of Pat Galloway

Construction projects don't always go as planned. When disputes arise, civil engineer Patricia Galloway (b.1957) is called in to resolve them fairly. She handles advanced construction projects such as a bridge in Hong Kong, airports in Malaysia, a refinery in Saudi Arabia and schools, hospitals, manufacturing plants, and sports arenas around the world. Pat even works in Australia part of the year. She is one jet-setting engineer!

PAT GALLOWAY is the past president (2003–2004) of the American Society of Civil Engineers, the first woman to hold the position in the Society's 152-year history.

Travel is a big part of Pat's life outside work, too. She regularly visits the New Jersey winery she owns with her husband, Kris Nielsen, and the two take an annual adventure vacation. They've visited Angkor Wat in Cambodia, Machu Picchu in Peru, and Africa, Antarctica, and the North Pole, among other destinations.

As the CEO of Nielsen-Wurster Group, Pat travels around the world to analyze schedule delays, damages, unforeseen costs, and other issues that can bring construction work to a halt. "All sides want an independent, third party to render a decision based on the facts," says Pat. "That's why they hire me."

From her home base east of Seattle, Pat and her border collie, Rings, board a corporate jet headed for an East Coast project. "Rings travels everywhere with me," says Pat. "Sometimes clients ask me to bring him into the conference, where he always sits obediently. Rings is well-received, and, as a result, people perceive me as someone who values my personal life as much as my professional life."

At left, Pat reviews a construction contract before a meeting. "I was hired to determine if a building contractor had a legitimate reason to delay construction," she says. "It may be tense during the negotiations, but when the decision is made, everyone breathes a sigh of relief because they can move on to other goals and projects."

After a business meeting, Pat may address a local professional group, such as the Society of Women Engineers. "When I speak, especially to women, I often repeat what my mother told me: 'Don't let anyone tell you it can't be done. You can do anything if you put your mind to it.'"

At right, Pat's ranch home near the Cascade Mountains offers relaxation on weekends. "I take trail rides whenever I can."

"If you don't take a vacation, your mind isn't as sharp and your body isn't rested," says Pat. At far left, Pat stands on the Great Wall of China. At near left, Pat explores Greenland.

"It's important for girls to understand they can be successful and have a fulfilling personal life," says Pat. "My husband Kris is the wind beneath my wings. We're a real partnership." At right, Pat and Kris at an American Society of Civil Engineers dinner.

Engineering and Law: A Perfect Fit

There are nearly as many branches of law as there are types of engineering. But there are several specialties in both fields that overlap. Thus, an engineering education becomes an ideal springboard to practice law.

For example, a patent lawyer helps an inventor obtain a patent, which is a government-issued document that tells the world that the inventor has full rights to his or her invention. The patent lawyer must fully understand the invention and the technical field in order to advise the inventor and draft a patent application. Engineering or a similar technical background is a prerequisite to practicing patent law.

Many environmental issues are both technical and legal, because regulations and laws govern how humans and businesses can affect the environment. In contract law, understanding how projects are designed and built—an ideal fit with engineering—helps greatly when contracts need to be written and enforced fairly.

Great Victory

Eleven women had pled cases to the U.S. Supreme Court before her, but in 1923 **FLORENCE KING** (1870–1924) was the first woman to win a Supreme Court case: "Crown vs. Nye," a patent infringement suit, which is still considered a precedent today. "Yes, it was a hard grind," she noted at the time. "I gave the best of my years to the career I had wanted since a tiny dot. But I won and I am happy."

Florence graduated from Kent College of Law in 1895 and then enrolled in night classes at the Armour Institute of Technology (now the Illinois Institute of Technology) to earn a degree in mechanical and electrical engineering. In 1897, she became the 685th person, and the first woman registered to practice before the U.S. Patent Office. Her stellar career was cut short in 1924 when she died of breast cancer.

The first woman patent attorney was also a mechanical and electrical engineer! Florence King was the first woman to win a case before the U.S. Supreme Court.

ENVIRONMENTAL ADVOCATE

"There are opportunities—large and small—for everyone to do something to protect and improve our environment," says civil engineer and attorney **CONNIE H. KING** (b.1953). "However, it just so happens that many environmental issues are both technical and legal."

That's why Connie's environmental engineering background is such an incredibly valuable asset to her law practice and her clients. Connie went into environmental engineering because she feels that if you help the environment, you also help people.

She emphasizes that her knowledge of engineering and law enables her to look at problems in a broader way. She's then able to come up with interesting solutions that other people might not consider.

Connie tells the story of a landfill in Colorado Springs that contaminated surrounding properties. Many years back, a judge ruled that the landfill owners didn't have to clean up the contamination.

However, when it came time for the owners to sell some adjacent property, Connie helped bring together a team of state and local government officials, attorneys, and technical people to make sure the landfill owners funded the clean-up of their mess before selling the property. Connie's team won, and the clean-up has been completed!

CONNIE KING PRACTICED ENVIRONMENTAL ENGINEERING FOR 13 YEARS WITH EXXON AND EASTMAN-KODAK. IN 1989 SHE WAS ADMITTED TO THE COLORADO STATE BAR AND HAS BEEN PRACTICING LAW EVER SINCE—DESPITE UNDERGOING TWO BOUTS OF BREAST CANCER. "ALTHOUGH I WORKED FEWER HOURS, MY WORK WAS A GOOD DISTRACTION FROM THE SIDE-EFFECTS OF CHEMOTHERAPY," SHE SAYS.

AN INTRIGUING PROFESSION

In the engineering world, two plus two always equals four, explains mechanical engineer and patent attorney **JANDA' CARTER** (b.1965). "This appeals to my objective and rational side. But with patent law, there are so many variables that appeal to my creative side."

As an engineer, Janda' worked at General Motors in the Buick, Oldsmobile, and Cadillac divisions. She then moved into new food processing technology at Quaker Oats. There, she met a corporate attorney. "Law was new. It was intriguing. I got hooked."

Janda' now represents clients with a limitless breadth of inventions. She has to understand each one. "I love working with the inventions—and with the inventors. They've worked hard to come up with something exciting and new. I help them secure the recognition they deserve."

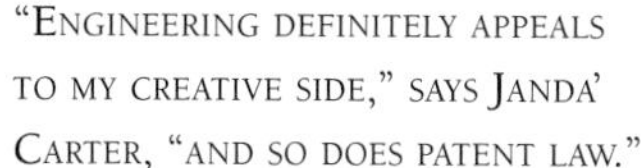

"ENGINEERING DEFINITELY APPEALS TO MY CREATIVE SIDE," SAYS JANDA' CARTER, "AND SO DOES PATENT LAW."

The Engineer Ambassador: Waging Peace

In honor of the new millennium, the United Nations issued its Millennium Development Goals, demonstrating its commitment to freeing the entire human race from want. The eight goals include: (1) eradicate extreme poverty and hunger, (2) achieve universal primary education, (3) promote gender equality and empower women, (4) reduce child mortality, (5) improve maternal health, (6) combat HIV/AIDS, malaria, and other diseases, (7) ensure environmental sustainability, and (8) develop a global partnership for development.

Engineers, being a central element of human society, are in a unique position to make significant progress toward these goals!

Lilia Abron wanted clean, bright homes for South African families, so her company, PEER, helps build energy-efficient houses that cost about $3,500. Here Lilia (far right) is pictured with new homeowners. She's now helping residents get energy-efficient appliances for cooking and heating their new houses.

Energy-Saving Homes

In 1994, when South Africa ended its enforced racial segregation—called apartheid—millions of people needed homes.

"Private contractors started building little more than concrete shacks as a move toward formal housing. It took a woman to say, 'that's not right'," says chemical engineer **LILIA A. ABRON** (b.1945). "Everybody deserves decent housing, especially families." She knows first-hand as the mother of three sons.

Lilia's company, PEER, went to work with an architect designing affordable, environmentally friendly, energy-efficient houses. "We developed the science inside the house, including a thermal insulation design, windows for solar heating, and solar hot water heaters." Local contractors were then able to build the houses.

"I call it 'compassionate engineering'," says Lilia. "Engineers have the skills to work the right way. They just need to remember that people's needs are an integral part of the project."

In 1999, Lilia won an award from the United Nations for helping to solve the housing problem without negatively affecting the environment. Her homes use 50 to 60 percent less energy than traditional houses.

THE ROAD OUT OF POVERTY

One of the many wrongdoings committed by the former apartheid regime of South Africa was a systematic denial of a decent education to non-white South Africans. In fact, the Republic of South Africa banned arithmetic from non-white schools in 1947 and didn't restore it until the late 1980s!

Civil engineer **GLORIA JEFF** (b.1952) was the associate administrator of the U.S. Federal Highway Administration in 1993 when the apartheid system crumbled. President Bill Clinton asked Gloria to head a technology transfer program with South Africa designed to help South Africans learn to build roads into the isolated interior of the country.

"In the U.S., we use sophisticated equipment and technology to design and construct roads with the least amount of costly labor," Gloria explains. "In South Africa, however, our goal was to put as many to work as possible. In an economy where there was 75 percent unemployment within the black community, our first priority was to give people meaningful jobs."

Gloria tells about one community project, a "two-hitter" she calls it. A town needed new roads. There was a brick-making factory in town, but not much demand for brick. So they trained the townspeople to build brick roads. This put the brick-makers—and the road-builders—to work

LOW-TECH, HIGH-IMPACT

By day, civil engineer **CATHY LESLIE** (b.1961) manages a staff of 30 planners and engineers in Denver who design roads, pipelines, and stormwater projects. At night, as executive director of Engineers Without Borders-USA, Cathy spends three to five hours managing an ever-growing number of volunteers—half of whom are women.

Cathy, who loves to travel, began her work in developing countries as a Peace Corps volunteer in Nepal, where she developed solutions related to drinking water and sanitation projects. Today EWB-USA partners with developing communities around the world to improve their quality of life.

EWB-USA believes in creating environmentally and economically sustainable engineering projects. It also believes in creating internationally responsible engineering students. "The villages that partner with EWB-USA put their hope in us," says Cathy. "I get a great deal of satisfaction knowing that we are making the children's lives better than their parents'."

ABOVE: CATHY LESLIE (SEATED, LEFT) AND COLLEGE STUDENTS FROM THE UNIVERSITY OF COLORADO AT BOULDER, PROFESSIONALS, TRANSLATORS, COOKS, AND OTHERS JOINED WITH VILLAGERS FROM FOUTAKA ZAMBOUGOU, MALI, TO BRING MUCH-NEEDED CLEAN WATER AND SANITATION TO THE VILLAGE. IT WAS CATHY'S VERY FIRST PROJECT WITH EWB-USA.

AT RIGHT: IMAGINE THE DELIGHT IN SEEING A PHOTOGRAPH OF AN OCEAN AND BREACHING WHALE FOR THE FIRST TIME. EWB-USA COORDINATED THE EFFORTS OF STUDENTS AT THE UNIVERSITY OF COLORADO AT BOULDER IN COOPERATION WITH THE HIMALAYAN LIGHT FOUNDATION TO INSTALL A SOLAR-POWERED COMPUTER IN KOT-TIMAL, A HIMALAYAN VILLAGE AT 12,000 FEET.

FAR LEFT: AS THE DIRECTOR OF THE MICHIGAN DEPARTMENT OF TRANSPORTATION, GLORIA JEFF HAS PLENTY OF OPPORTUNITIES TO TALK TO PEOPLE ABOUT TRANSPORTATION—AND ABOUT THE IMPORTANCE OF EQUAL OPPORTUNITY AND PREVENTING DISCRIMINATION.

LEFT: CHILDREN IN SOUTH AFRICAN SHANTY TOWN

STORMY WEATHER

In October 1998, "Mitch," a category five hurricane, with 180 m.p.h. winds, decimated much of Central America. Honduras and Nicaragua were especially hard hit. Thousands of people died, and millions were left homeless. Floods and mudslides ravaged the landscape.

Hydraulic engineer **SONIA MAASSEL JACOBSEN** (b.1955) joined other U.S. Department of Agriculture employees traveling to Nicaragua to assess damage and determine the aid needed. Despite crumbling roads and bridges, Sonia headed into areas so isolated she carried a global positioning system (GPS) device to track her location.

Sonia met with local farmers and nonprofit organizations to evaluate damages, recommend projects, and estimate costs. Many projects involved assessing the condition of roads, water systems, and other agricultural infrastructure that had been damaged by erosion, debris, sediment, and boulders. How could it be restored? Should the land be abandoned for farming?

Says Sonia, "My experience in Nicaragua was a remarkable adventure in my life. I'm happy I chose a career where I can work with farmers and landowners to protect the earth and provide food for the world."

ABOVE: SONIA JACOBSEN IS AN AGRICULTURAL ENGINEER WITH THE U.S. DEPARTMENT OF AGRICULTURE—NATURAL RESOURCES CONSERVATION SERVICE (NRCS) IN ST. PAUL, MINNESOTA.

ABOVE LEFT: IN THE AFTERMATH OF HURRICANE MITCH, SONIA FOUND MANY SMALL STREAMS AND CULVERTS IN NICARAGUA FILLED WITH DEBRIS, WHICH BLOCKED DRAINAGE AND THREATENED FARMLAND WITH MORE FLOODING WHEN THE NEXT RAINY SEASON BEGAN. "WAS THE BEST SOLUTION TO CLEAN OUT DEBRIS OR RE-ROUTE WATER? HOW COULD STREAMBANKS BE STABILIZED? THESE WERE SOME OF THE ISSUES I HAD TO CONFRONT," RECALLS SONIA.

WHEN DISASTER STRIKES

"Engineering is an essential skill to be able to meet basic human needs," says civil engineer **JO DA SILVA** (b.1967). "Oddly, in the past, relief agencies did not recognize the importance of engineers, so it was difficult for many to identify and recruit engineers after a natural or humanitarian disaster." Registered Engineers for Disaster Relief (RedR), was founded to create a register of engineers who could be called on at short notice to work with front-line relief agencies.

Jo is a founding member of RedR-International. Her first assignment was to provide humanitarian relief to the massive exodus of people from Rwanda into Tanzania in 1994, after an estimated 800,000 Rwandans were killed. A quarter of a million people flocked into the village of Ngara, Tanzania. Jo employed up to 200 refugees a day to set up huge camps from scratch—things as basic as shelter, water, latrines, and first-aid stations.

JO DA SILVA SPENT SEVEN MONTHS IN SRI LANKA OVERSEEING THE CONSTRUCTION OF 55,000 SHELTERS TO PROVIDE HOUSING FOR THE HALF-MILLION PEOPLE DISPLACED BY THE DEVASTATING TSUNAMI THAT HIT THE COASTAL REGIONS IN LATE DECEMBER 2004.

Tikun Olam

In the Jewish tradition, Tikun Olam means "repairing the world," making the world a better place for all. It is one of industrial engineer **SHULAMITH "SHULA" KOENIG'S** guiding principles, which she first put to use in Israel designing and manufacturing water-saving devices for irrigation.

"Water—just as human rights—means life, growth, health, and well-being," says Shula. She describes human rights as the banks of a river in which life can flow free. Human rights—learning the "rules of the game"—strengthen those banks. Within those banks, every drop of water has value. If you add a drop, the river will run faster. Take one drop out, the river will flow slower. Take too many drops out, and the river will dry up. Human rights defines the value of every woman, man, and child. In the river of life, people must know human rights to transform the world."

In 2003, Shula won the prestigious 2003 United Nations Award for Outstanding Achievement in the field of Human Rights for her work with PDHRE (the People's Movement for Human Rights Education), an organization she founded to energize, organize, and facilitate learning about human rights, and make human rights education relevant to people's daily lives. In this, Shula says, she is fulfilling her commitment to Tikun Olam, one drop at a time.

YUVA (Youth for Unity and Voluntary Action) works on the issues of housing and livelihood for marginalized women, children, and youth in urban and rural areas of India. Shula Koenig (above, far right) attended the inauguration of the YUVA Center in Mumbai, India. At left, Shula (second from the right) joins community leaders from Mali in the inauguration of PDHRE Africa in Mali.

Irrigation 'Round the Globe

Agricultural engineer **DOROTA ZOFIA HAMAN** (b.1951), an expert in irrigation water management, has traveled throughout the world helping farmers learn more about growing healthy food by using various methods of irrigation.

Many non-governmental organizations (NGOs) distribute simple drip irrigation kits that contain plastic pipe that connects to a water source; farmers operate them by a foot treadle pump to circulate water. While these kits are simple, drip irrigation can be very complex. That's where Dorota excels in helping farmers operate their systems more efficiently.

"By happy accident, I fell into my career while reviewing some interesting data during a math project. Now I find tremendous reward in helping people worldwide grow more and better food while putting less stress on the environment."

In Zimbabwe, Dorota Haman joined a team from the United Nations Food & Agricultural Organization to teach small farmers how to make the most of their water supplies. Her course drew students from all over South and East Africa who wanted to learn about drip irrigation. Here, Dorota is shown operating a treadle pump used for irrigation.

YOU CAN MAKE A DIFFERENCE!

We live in an exciting, complex world where the need for women's creativity and know-how has never been greater. Why? Because women hold unique insights into what's important in this world. And women have a unique approach to solving problems.

One viewpoint or one way of thinking is no longer adequate. Engineering needs women—their hearts, their minds, their team spirit, and their good sense!

The stories of the accomplishments of women engineers in this book are amazing. But keep in mind they are the accomplishments of real women. Not superhuman women. Not nerdy women. Not women content to work alone.

But women with families. Women who pursue their passion. Women who earn good incomes. Women who travel the world. Women who live fun, fulfilling lives at work and at home.

Can one woman change the world? What about many women armed with powerful engineering skills? The women in this book say, "Yes!"

See if you can see yourself standing in the shoes of an engineer someday!

So Many Options

Engineering offers so much variety and so many options that it may be difficult to choose just one area for a career. But imagine the possibilities! There will surely be a path for you. How to decide? Here are some sage words of wisdom for those thinking about a career in engineering—from women engineers who have been there.

On Following Your Heart:

Mary Leigh Wolfe tells her students, "Stay open-minded. The diversity in engineering careers is amazing, even mind-boggling. Just because you haven't heard of it doesn't mean it's not out there. Keeping looking for what you want to do."

"Engineering is such a broad career that it can create the kind of opportunities to keep you energized for a lifetime," says **Suzanne Jenniches**. "You don't have to be brainy, you just have to be interested and then find the right kind of engineering for you."

Maryly Van Leer Peck concurs. "Don't close that door! Take advantage of every opportunity."

On Being Prepared:

"Bill Ford [the CEO of Ford Motor Company and great-grandson of Henry Ford] once told me that the only thing he regrets in his life was not focusing on math and science," says **Susan Cischke**. "Today, I tell girls, 'Take as many math and science classes as you can—even in middle school.' Developing technical and logic skills early on is a tremendous advantage."

Although **Cathy Leslie** emphasizes that having math and science skills is important, she says, "Don't let math and science scare you. There's more to engineering than just bookwork. Engineering is just as much about people and cultures as it is about math and science."

Erin Wallace agrees. "Engineering is so much more than the math. Girls shouldn't be discouraged if they're not whizzes with numbers. It's about problem solving and pushing yourself to find creative solutions."

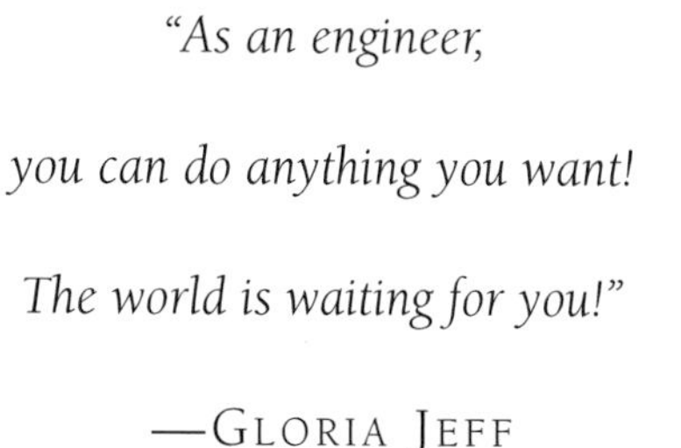

"As an engineer, you can do anything you want! The world is waiting for you!"

—Gloria Jeff

Alma Martinez Fallon says, "Don't let anyone tell you that you can't do it! You can . . . and you'll be great at it! Surround yourself with good friends and colleagues who will support you along the way."

On Your Career:

Rachel Thomas provides some good advice on choosing a college. "Research various fields of engineering first, then choose a college that offers the best program. Discovering the advantages and disadvantages are important, so ask questions!"

Diane Morse emphasizes that field experience is the key to advancement for some engineering specialties. But she admonishes (with a twinkle in her eye), "Never buy the steel-toed tennis shoes for your field work. No one will take you seriously. Get the standard ugly brown leather steel-toed boots and you'll look like you belong."

Priscilla Nelson describes her engineering career as non-linear, meaning that she didn't follow the typical path to becoming an engineer. "We change directions and have many careers during our lifetimes," she says. "No decision you make when you're young cannot be unmade."

Acknowledgements

Engineers, as a rule, are a modest bunch, often letting their work speak for themselves. The problem is that, in the unaccustomed eyes of the public, engineers' work is barely discernable. Most people just don't realize the profound impact that engineers and engineers' work has on our day-to-day quality of life.

An even larger problem is the lack of recognition for women engineers' remarkable accomplishments. If the work of engineers is subtle, the work of women engineers has remained largely invisible!

The purpose of *Changing Our World: True Stories of Women Engineers* is to acknowledge —and bring to light—the countless contributions made by women engineers to bettering our world. And the purpose of this acknowledgement is to pay tribute to the many people behind the scenes who brought this book to fruition.

Patricia Galloway, Ph.D., P.E., the first woman president of the American Society of Civil Engineers, was the catalyst for this book. Pat envisioned a book that would tell the stories of past and present women engineers who could serve as role models to girls—and then, through diligent and indefatigable efforts, brought together the resources to make it happen.

Jane Howell served as the chair for the project team that steered this project to completion. Jane's creativity and leadership have been a true inspiration. Similarly, Susan Skemp, through her role as the chair for the project's advisory committee, has made many thoughtful suggestions and provided sage direction on the book and the project overall.

Sincere thanks goes to Martha Moore Trescott for her voluminous research on women engineers. Her extensive and meticulously detailed volume, *New Images, New Paths: A History of Women in Engineering in the United States 1850–1980* is an invaluable repository of the fascinating and richly varied lives of women engineers, as well as the history of their hard-earned advances over 130 years. Similar thanks goes to Deborah Rice, archivist for the Society of Women Engineers, who provided extensive information on many of the women featured in this book.

Kendal Andersen, Vicki Speed, and Suzanne Storar were of tremendous assistance in researching, interviewing, and writing about the women in this book. Lisa Elliot, the graphic artist, gave the book life. Patrick J. Natale, P.E., Jeanne Jacob, and the ASCE Foundation staff raised the money to produce the book. Suzanne Coladonato, manager of book production, and Ann Pallasch, publicist, brought it to press.

Michael Geselowitz, Peggy Layne, José Medina, Patricia Paddock, and Jill Tietjen were all kind enough to review drafts of the book, and Sheila Chandrasekhar edited it.

The Extraordinary Women Engineers Project would not be possible without the generous contributions of many women and men (see page 216) who believe that the time is now for bringing more girls and women into the incredibly rewarding field of engineering. We are grateful for the generous leadership gifts received from the American Society of Civil Engineers, Stephen D. Bechtel, Jr., National Council of Examiners for Engineering and Surveying, National Science Foundation, Tyco Electronics Corporation, and United Engineering Foundation, all of which provided the money to launch the project.

Above all, however, is a heartfelt acknowledgement of the time and effort that the women engineers featured in this book—and the many, many others whose stories are equally compelling—have dedicated to their careers, their lives, their families, and to making the world a better place for us all.

—Sybil E. Hatch, P.E.
Berkeley, California

Contributors

The Extraordinary Women Engineers Project is the first large-scale project jointly conceived of and undertaken by a multi-disciplinary group of engineering societies. The entire project team is extremely grateful for the generous support provided by the individuals, corporations, universities, and associations listed below. These gifts stand as a true testimonial of the deep conviction that the donors all share: that engineering is an extremely rewarding career and that a diversified workplace is vital to the future of engineering.

TITANIUM LEVEL $100,000 & Greater
Stephen D. Bechtel, Jr. • United Engineering Foundation, Inc.

PLATINUM LEVEL $50,000–$99,999
American Society of Civil Engineers

GOLD LEVEL $25,000–$49,999
Tyco Electronics Corporation

SILVER LEVEL $10,000–$24,999
National Council of Examiners for Engineering and Surveying

COPPER LEVEL $5,000–$9,999
Patricia D. Galloway, Ph.D., P.E., F.ASCE • WTS International

BRONZE LEVEL $1,000–$2,499
Jean C. Carter • Jennie Lee Colosi, P.E. • Katherine M. Fitzpatrick Donohue • Michael D. Kennedy, P.E. —CH2M HILL • M. Kathryn Knowles, Ph.D. • Joyce and Walter LeFevre • Rockwell Collins • Boston Society of Civil Engineers • American Society of Heating, Refrigerating and Air-Conditioning Engineers, Inc.

BRASS LEVEL $500–$999
Michelle J. Anschutz, P.E. • Marsha Anderson Bomar—Street Smarts • Bente Hoffmann Eegholm Sarah E. Hancock • Sybil E. Hatch, P.E. • Agnes Sims Hite • Brandy E. Iglesias • Carl and Maria Lehman • Yvonne R. Lyda • Kathleen H. McCauley, P.E. • Elizabeth A. McMillan • Catherine A. Morris, S.E. • Patrick J. Natale, P.E., and Sheila Natale • Prof. Margaret S. Petersen • Jamee Sue Plockmeyer, P.E. • Saw-Teen See, P.E., Hon.M.ASCE • Megan Marie Siarkiewicz • Charlotte Tyson • Dr. Jessica Winter • American Institute of Aeronautics and Astronautics • Castle Contracting, LLC • Hayden Consultants, Inc. • PBS&J

SUPPORTER LEVEL $250–$499
Anonymous • Anni H. Autio, P.E. • Debra Bogdanoff • Janet Petra Bonnema, M.S.C.E. • Denise A. Bunte-Bisnett, P.E. and Rachel V. Bisnett • Anna Lankford Burwash • Shirley E. Clark, Ph.D., P.E., • Barbara J. Colony, P.E. • Kathy J. Caldwell, P.E. • Wendy B. Cowan • Margaret A. Curtis, P.E. • Marla Dalton, P.E. • Teresa Lamb Dellies, P.E. • In Memory of the Honorable Aeronautical, Scientist, Engineer, Vladimir Deriugin, Sr.—an American Hero • Casey Dinges • John E. Durrant, P.E. • Carol A. Ellinger, P.E. • Dr. Elahe Enssani, P.E. • Sharon L. Estes • Dr. Thelma Estrin • Meggan Farrell • Luiz Felipe and Alicia Figueiredo • Barbara G. Fox • Mary Moloseau Goetz, P.E. • Margaret and Michael Goode • Bruce and Deborah Gossett • Janine L. Grauvogl-Graham, P.E. • Roger and Charlene Helgoth • William P. Henry, P.E., F.ASCE • Elise Hosten-McGough • Sandra L. Houston, Ph.D. • Jane Howell • Stacy Lewis Hutchinson • Beatrice Stein Isaacs, Ph.D., P.E. • Jeanne G. Jacob and Richard G. Frank, In Memory of Clara and Alfred Jacob • Shannon H. Jordan, P.E. • Robin A. Kemper, P.E., F.ASCE and Family • Nathelyne A. Kennedy, P.E. • Stephanie K. Kinsey • Richard and Kathy Klein • Teresa Kulesza • Ha Thu Le • Deborah H. Lee, P.E., P.H. • Mr. and Mrs. Mark Leeman • Dr. and Mrs. Thomas A. Lenox • Neville S. Long, P.E., F.ASCE • John V. Lowney, P.E. • Marnita A. Magner • Elizabeth McDargh • Catherine McMullen, P.E. • Carolyn J. Merry • Julie Jordan Metts, P.E., M.B.A. • Barbara and Brian Minsker • David and Janet Mongan • Patricia A. Montgomery • Diane I. Morse • Jon L. Muller, P.E. • Judith Nitsch, P.E. • Dr. Debra R. Reinhart, P.E. • Lawrence H. and Gail W. Roth • Dr. and Mrs. Jeffrey S. Russell • Grecia R. Matos and Michael R. Sanio • Vicki Scharnhorst, P.E. • Betty Shanahan • Teresa Herman Shea • Robert Silverstein • Tom and Marcia Smith • Kate Smith, P.E.—Trillium Consulting • Akiko Y. Sulisufaj • Karolyn Thompson • Kanathipillai Vijayaratnam • Roli and James Wendorf • Lisa Wieland Larson, P.E. • Emily E. Wieringa • Lee L. Wolfe, P.E. • Mark Woodson—Woodson Engineering and Surveying, Inc. • Theresa Wynn • Jin Zhang • Department of Civil and Environmental Engineering —University of Wisconsin-Madison • Liftech Consultants Inc. • Women in Engineering Programs & Advocates Network (WEPAN), Inc.

Extraordinary Women Engineers Project Coalition

The Extraordinary Women Engineers Project is an unprecedented awareness and outreach program designed to encourage young women to choose engineering as a career and to develop a new generation of role models for those already in the field. The project has been endorsed and supported by each member of the Extraordinary Women Engineers Project Coalition:

Accreditation Board for Engineering and Technology, Inc

Alliance of Technology and Women

American Association for the Advancement of Science

American Association of Engineering Societies

American Council of Engineering Companies

American Institute of Aeronautics and Astronautics

American Institute of Chemical Engineers

American Institute of Constructors

American Physical Society

American Public Works Association

American Society of Agricultural Engineers

American Society of Certified Engineering Technicians

American Society of Civil Engineers

American Society of Mechanical Engineers

American Society of Naval Engineers

American Society of Plumbing Engineers

American Water Works Association

Arizona State University, Ira A. Fulton School of Engineering

AACE International—The Association for the Advancement of Cost Engineering

ASFE

Biomedical Engineering Society

Construction Management Association of America

Colorado School of Mines, Division of Engineering

Illuminating Engineering Society of North America

Institute of Electrical and Electronics Engineers

Institute of Industrial Engineers

Japan Society of Civil Engineers

Junior Engineering Technical Society

Kettering University

MentorNet

Mississippi State University, James Worth Bagley College of Engineering

National Academy of Building Inspection Engineers

National Academy of Engineering

National Academy of Forensic Engineers

National Association for College Admission Counseling

National Association of Women in Construction

National Council of Examiners for Engineering and Surveying

National Engineers Week Foundation

National Institute of Ceramic Engineers

National Society of Black Engineers

National Society of Professional Engineers

North Carolina State University, College of Engineering

Purdue University, College of Engineering

Rowan University

Society of American Military Engineers

Society of Automotive Engineers, Women Engineers Committee

Society of Fire Protection Engineers

Society of Naval Architects and Marine Engineers

Society of Women Engineers

TMS—The Minerals, Metals, and Materials Society

Texas Tech University

University of Delaware, College of Engineering

University of Texas, San Antonio College of Engineering

University of Wisconsin—Madison

Women in Engineering Programs & Advocates Network

Women's Transportation Seminar

WGBH

Extraordinary Women Engineers

Although the number is hard to pin down, estimates indicate that there are over a million engineers practicing in the U.S. today. Around 100,000 are women. And each one has her own extraordinary story to tell.

Only a small portion of their stories—238 to be exact—could be included in this book. But there were many more women engineers nominated! The engineering community thinks highly of these women—rightfully so!

Perhaps you'll meet a woman engineer one day, or listen to her talk about her career, or watch her on T.V. Perhaps some day you'll become a woman engineer. And then someday, instead of 100,000 women engineers in this country, there will be half a million or more! Now wouldn't that be something?

Women profiled in the book:

Women nominated but not included in the book: Janet Adams • Natalie J. Adelman • Adjo Amekudzi • Maria Carmen Andrade • Lisa J. Arganbright • Ethel H. Bailey • Julie Bailey • Kimberly K. Baxter • Emily T. Bellenger • Vaijayanti Bendre • Sophia Hayden Bennett • Carla Birk • Susan M. Blanchard • Alexandria Boehm • Sandra Bouckley • Gail Boydston • Elizabeth Bragg • Patricia L. Brown • Jeanne Harris Bruck • Kathy J. Caldwell • Jackie Carpenter • Julie Carter • Rachel Chandler • Sara Heger Christopherson • Yvonne Y. Clark • Cyn Coleman • Mary Ellen Collentine • Isabel Coman • Adelaide Cooper • Margaret Arronet Corbin • Iris Cummins • Amber Miller Cutcliff • Suzanna Darcy-Hennemann • Semahat S. Demir • Lisa DeSantis • Kelly Detra • Carol J. Devoy • Catherine M. Downen • Mildred S. Dresselhaus • Merette Elsayed • Olive E. Frank • Minette Ethelma Frankenberger • Jane Frankenberger • Isabelle F. French • Debora Fronczak • Diana Gale • M. Elsa Gardner • Dixie Garr • Alice C. Goff • Martha Goodway • Lois Graham • Edith Griswald • Vinta Gupta • Annette Haddad • Emily Hahn • Dorothy Hall • Nancy Hamilton • Catherine Cleveland Harelson • Katheryn Hatcher • Rachel Heiner • Janet Hering • Nhuy Hoang • Katherine Hopper • Stacy L. Hutchinson • Mary Ellen Hynes • Helen Innes • Mary Kay Jackson • Rebeca E. Jemenez • Florence Hazel Caldwell Jones • Lucile B. Kaufman • Marshall Keiser • Nathelyne Archie Kennedy • Rachel Kerestes • Rita Klees • Lisa C. Klein • Stavroula Kolitsopoulos • Blyth S. Koyiyemptua • Kahne Krause • Jennifer E. Krutz • Suzanne Lacasse • Sandra Larson • Margaret Law • Duy-Loan Le • Laura Sables Leber • Eva K. Lee • Kathy Leo • Suzanne LeViseur • Michele Lezama • Terri Love • Linda C. Lucas • Marie E. Luhring • Amy Lynch • Mary E. Lynch • Jo Ann Macrina • Nancy J. Manley • Cynthia Bergman Manning • Anne Marselis • Frances Marshall • Norma Jean Mattei • Shirley McCarty • Maureen A. McDonough • Carrie McElwein • Elizabeth Jackson McLean • Rachel A. B. McQuillen • Teresa H. Meng • Elizabeth A. Hood Messer • Borjana Mikic • Mary Thelma Miller • Kimberly A. Moore • Mary L. Murphy • Claudia Myers • Loring Nicholson • Gabriele G. Niederauer • Elizabeth Oolman • Stacia L. Palser • Mildred Paret • Marion Sarah Parker • Shayn Peirce • Celina Ugarte Penalba • Helen Joyce Peters • Mildred Pfister • Elaine R. Pitts • Lila Poonawalla • Julie Portelance • Patti Psaris • Hazel Irene Quick • Marie Reith • Rebecca Richards-Kortum • Harriet B. Rigas • Mabel MacFerran Rockwell • Nellie Scott Rogers • Lori J. Ryerkerk • Diane A. Schaub • Marlene Schmidt • Cheryl B. Schrader • Susan Selke • Ruth I. Shafer • Holly Shill • Anna Shmukler • Amy Smith • Margie Smith • Mary Olga Soroka • Carol Stadler • Anna G. Stefanopoulou • Anne Steinemann • Rachel Stender • Stacey Stenerson • Lori A. Stetton • Sherri K. Steuewer • Martha Dicks Stevens • Margaret R. Taber • Mary Ann Tavery • Valentina Tereshkova • Anita Thompson • Fay Collier Tresidder • Tana L. Utley • Jean S. Vander Gheynst • Sophie Vandebroek • T. Kyle Vanderlick • Elisabeth Vanzura • Barbara Vassalluzzo • Sara Wadia -Fascetti • Oksana Wall • Catherine E. Walshe • Naomi L. Weathers • Yvette P. Weatherton • Josephine R. Webb • Frankie Barnett Welsh • Jane Wernick • Andrea Wesser • Nancy Wheeler-Nelson • Doris Willmer • Rhonda Wilson • Lisa Woods • Margaret Wooldridge • Yoon Song Yee • Ruihong Zhang • Mary Ann Zimmerman

Photo Credits

Unless otherwise specified, photographs were provided by the subject.

Cover Clockwise from top left: State of California Department of Public Works Division of Highways District VII; Hyoungshin Park; John Livzey Photography; NASA; Leslie E. Robertson Associates; University of Kentucky; © Felice Macera; Adam D. Mattivi **vii** Courtesy of NASA **viii** Left: SWE Archives, Walter P. Reuther Library, Wayne State University **ix** Bottom: James Kegley Photography **4** © Royalty-Free / Corbis **5** Left: © Royalty-Free / Corbis; right bottom: John Livzey **6** Left bottom: Randy Montoya, Sandia National Laboratories; right bottom: Jerome Brown **7** Left: Courtesy of ExxonMobil; right center: © Roger W. Winstead, North Carolina State University **8** Top: © Ryan McVay / Getty Images; bottom right: © Yellow Dog Productions / Getty Images; bottom left: © Lawrence Berkeley National Library / Getty Images **9** Top: © Patrick Sheandell O'Carroll / Getty Images; bottom left: © David Becker / Getty Images; bottom right: © UHB Trust / Getty Images **10** SWE Archives, Walter P. Reuther Library, Wayne State University **11** Top left: © James Cavallini / Photo Researchers, Inc.; top right: © Felice Macera; bottom: © Jim Wehtje / Getty Images **12** © Jim Wehtje / Getty Images **13** Bottom left: Bill Blanchard; bottom right: Zimmer® Hip System **14** Top: Courtesy of Cornell University Photography; bottom: Jacqueline Cole **16** Left: Frank Anderson; right: Jason Rife **17** Robert Hubner, Washington State University Photo Services **18** Left: courtesy of Medtronic, Inc.; right: Dan Johnson, Marquette University **19** Top: Hyoungshin Park; center left: Milica Radisic; center right: Helen Shing; bottom: © Comstock **20** Left: © JupiterImages; right: Paul Braly, Medical Illustrations, UNC SOM **21** Bottom: © Jennie Woodcock; Reflections Photolibrary / Corbis **22** Background: © Duncan Smith / Getty Images **24** Left: © Stockbyte / Getty Images; right: © Lawrence Lawry / Getty Images **25** Top: Jerome Brown **26** Left: Courtesy of the USDA Natural Resources Conservation Service; right: © David Young-Wolff / Getty Images **27** Top: © L. Lefkovitz / Getty Images; bottom: © Chris Windsor / Getty Images **28** Top: © Brian Hagiwara / Getty Images **29** Bottom: Courtesy of Alcoa, Inc. **31** Bottom: Prather Warren, LSU University Relations **32** Background: © CDC / PHIL / Corbis; inset: from the Collections of the University of Pennsylvania Archives **33** Top left: courtesy of the Herb and Dorothy McLaughlin Photograph Collection, Arizona State University Libraries; top right: Arizona Historical Society; bottom: © RML / Getty Images **34** Michael Freed, Loonar Photography **37** Top: Timothy F. Wheeler; bottom right: © Martin Ruegner / Getty Images **38** Left: © Darrell Gulin / Corbis **39** Top: Charles M. Brown **40** Top: © Ryan McVay / Getty Images; bottom left: © Corbis; bottom right: © Rodney Hyett, Elizabeth Whiting & Associates / Corbis **41** Top: © Petrified Collection / Getty Images; middle: © Siede Preis / Getty Images; bottom: © John Edwards / Getty Images **42** Left and middle: Leslie E. Robertson Associates; right: Kohn Pedersen Fox Associates **43** Leslie E. Robertson Associates **44** Top left: Bison Archives / Marc Wanamaker; top right: © Joseph Sohm; ChromoSohm Inc. / Corbis; bottom: Julia Morgan Collection, Special Collections, California Polytechnic State University **45** Top: courtesy of the University of Colorado at Boulder Archives; bottom: SWE Archives, Walter P. Reuther Library, Wayne State University **46** Left: © Bettmann / Corbis; middle: Schenectady Museum, Hall of Electrical History Foundation / Corbis; right: © Archive Photos / Getty Images **47** Top left: provided by Purdue University with permission of Ernestine Gilbreth Carey; top right: SWE Archives, Walter P. Reuther Library, Wayne State University; bottom: National Museum of American History, Smithsonian Institution **48** Left top: Jill S. Tietjen, P.E.; left bottom: SWE Archives, Walter P. Reuther Library, Wayne State University; Right top: © Bettmann / Corbis; right bottom: Kevin Rector **49** Top left and right: courtesy of Carrier Corporation Archive; bottom: Special Collections and Archives, University of Kentucky Libraries **50** Top: SWE Archives, Walter P. Reuther Library, Wayne State University; bottom left: © Image Source / Getty Images; bottom right: © Don Farrall / Getty Images **51** Bottom left: Nancy Hribar, Zena Photography **52** Top: provided by the City of Chicago Department of Water Management **53** Bottom right: © Connolly Steve / Corbis Sygma **54** Jim Cole **55** Top left and right, Jim Cole **56** Background: © Corbis; left and right: courtesy of MIT Museum **57** Top left: SWE Archives, Walter P. Reuther Library, Wayne State University; bottom left: Lisa Brothers, P.E.; bottom right: Thurmon Smith **58** Left: © Robert Glusic / Getty Images; right: © Bruce Heinemann / Getty Images **59** © Bruce Heinemann / Getty Images **60** Left: © Randy Wells / Getty Images; right: © Frank Oberle / Getty Images **61** Top right: Bill Johnson / U.S. Army Corps of Engineers **62** Top: SWE Archives, Walter P. Reuther Library, Wayne State Univ-ersity; bottom: Patrick M. Lynch **63** Bottom: Erik Sampers / Getty Images **64** Left: © Corbis; right: courtesy of the U.S. Environmental Protection Agency **65** Bottom left: © Bettmann / Corbis; bottom right: © Galen Rowell / Corbis **66** Top right: © Alvis Upitis / Getty Images **67** Top right: SWE Archives, Walter P. Reuther Library, Wayne State University; bottom left: © Carl Pendle / Getty **68** Top: © Richard Laird / Getty Images; bottom: courtesy of Cornell University **69** Bottom left: © Reuters / Corbis **72** Left: courtesy of the U.S. Geological Survey; right: Library of Congress, Prints & Photographs Division, Detroit Publishing Company Collection; LC-D428-10429 L, LC-D428-10429 C, LC-D428-10429 R **73** Top right: SWE Archives, Walter P. Reuther Library, Wayne State University **74** Left: © Jose Fuste Raga / Corbis; right: © David Sailors / Corbis **75** Top left: © Ariel Skelley / Corbis; top right: © K.-H. Hänel / Corbis; bottom: © José Fuste Raga / zefa / Corbis **76–77** Ford Motor Company **78** Volvo Cars of North America, LLC **79** Left: © Thomas E. Bedford; right: SWE Archives, Walter P. Reuther Library, Wayne State University **80–81** Dorsey Patrick Photography **82** Left: keithwoodphotography.com; right: Terry Blackburn Photography **83** Top: SWE Archives, Walter P. Reuther Library, Wayne State University **84** Top left and right: courtesy of University of California, Riverside; bottom: SWE Archives, Walter P. Reuther Library, Wayne State University **85** Bottom right: International Truck & Engine Corporation **86** Background: © Chad Ehlers / Getty Images; inset: State of California Department of Public Works Division of Highways **87** Top left and right and bottom right: State of California Department of Public Works Division of Highways District VII **88** Top left and right: © Andrew Gunners / Getty Images; bottom left: Tillman Allen Greer, LLC **89** Bottom left: WMATA Photo **90** Background: © G.E. Kidder Smith / Corbis; bottom left: Library of Congress Prints & Photographs Division, POS-TH-KIR, no. 21; bottom right: Library of Congress Prints & Photographs Division, LC-USZ62-120162 **91** Top left and right: John Consoli; bottom left and right: FIGG Engineering **92** Bottom left: ©

Michael Dunning / Getty Images; bottom right: © Thinkstock / Getty Images **93** Left: courtesy of Cornell University; right: from the Collections of the B&O Railroad Museum **94** Top left: Lawrence Technological University Public Relations Office; top right: John Linden Photography **95** Top left: Hornagold & Hills; top right: Terry Chisholm; bottom: John Livzey Photography **96** Left: courtesy of MIT Museum; right: Northrop Grumman Newport News **97** Top left and bottom: John Livzey Photography **98** Left: Joel Styer; right: Antony Nagelmann / Getty Images **99** Top: © Allana Wesley White / Corbis; bottom: © Tom & Dee Ann McCarthy / Corbis **100** © The Walt Disney Company **101** Great Coasters International, Inc **102** Top left: Michael Dickter, Magnusson Klemencic Associates; top right: Lara Swimmer Photography; bottom: Ellen M. Banner / The Seattle Times, copyright 2005, used with permission **103** Bottom left: David Yuill; bottom right: © Tom Brakefield / Getty Images **104** Top: © Stockdisc Classic / Getty Images; bottom: © H. Winkler / A. B. / zefa / Corbis **105** Sam Odgen Photography **106** Top: Ron Torres; bottom right: Polly Pocket is registered, owned by, and used under license from Origin Products Limited **107** Jonathan Leaders **109** Top: Ben Lawson **112** Background: courtesy of XM Radio; bottom left: courtesy of Texas Instruments **114** © Archive Holdings Inc / Getty Images **115** Top: © Ariel Skelley / Getty Images; bottom: © H. Winkler / A. B. / zefa / Corbis **116** Top: © Ryan McVay / Getty Images; bottom: © B. Anthony Stewart / Getty Images **117** Bottom: © Bell Telephone Co. **118** Top: courtesy of Neil B. McGahee; bottom: © StockTrek / Getty Images **119** Left: SWE Archives, Walter P. Reuther Library, Wayne State University; right: courtesy of RCA AstroElectronics **120** Left, center, bottom: © Corbis; right: © Roy Morsch / Corbis **121** Top left and right: courtesy of Harvard University Archives **122** Top left: courtesy of the Computer History Museum **124** Background: © Chad Baker / Getty Images; bottom left and right: courtesy of Cassandra Johnson **126** Top: John Marchetti; bottom: courtesy of Motorola **127** University Photography, Duke University **128** Background: © Corbis; inset: © Mike Theiss / Jim Reed Photography / Corbis **129** Top: © Digital Vision / Getty Images; bottom: © Russell Illig / Getty Images **130** Background: © National Archives / Corbis; inset: Library of Congress, Prints & Photo-graphs Division, LC-USZ62-97270 **131** Right: SWE Archives, Walter P. Reuther Library, Wayne State Univ-ersity; left: The Ohio State University Archives **132** Top: Library of Congress, Prints & Photographs Divi-sion, pan 6a10948; center: Special Collections, Schaffer Library, Union College; bottom left: © American Society of Mechanical Engineers **133** Top: SWE Archives, Walter P. Reuther Library, Wayne State University; bottom: © Bechtel Corp **136** Library of Congress, Prints & Photographs Division, FSA / OWI Collection, left: LC-USW3-006673-D, center: LC-USZ62-130502, right: LC-USZ62-83180 **137** Top left: courtesy of MIT Museum; top right: SWE Archives, Walter P. Reuther Library, Wayne State University; bottom right: © Douglas Mesney / Corbis **138** Left: © Corbis; right: © Roger W. Winstead, North Carolina State University **140** Top: courtesy of the U.S. Army; bottom: © Stuart Westmorland / Corbis **141** Top: © Matthew McVay / Getty Images; bottom: courtesy of the U.S. Air Force **142** Courtesy of NASA **143** Top left: Jose Mercado / Stanford News Service; top right: Stanford News Service; bottom left: collection Dean Unger; bottom right: © Corbis **144** Courtesy of NASA **145** Top left and right: Katharine Stinson Collection, Special Collections Research Center, North Carolina State University Libraries; bottom: © 2005 Simon Griffiths **146** Top: SWE Archives, Walter P. Reuther Library, Wayne State University; bottom: TM & © Boeing, used under license **147** Left: courtesy of NASA; right: SWE Archives, Walter P. Reuther Library, Wayne State University **148** Left: Paul Maritz; right: SWE Archives, Walter P. Reuther Library, Wayne State University **149** Top: courtesy of NASA; bottom left: courtesy of Aviation Antiques; bottom right: Library of Congress, Prints & Photographs Division, LC-USZ62-119492 **150** Left: © Dean Conger / Corbis; right: © Andy Caulfield / Getty Images **151** Bottom: Bob Ferguson **152** Top left: courtesy of the U.S. Air Force; top right: courtesy of David Chan, NASA National Transonic Facility; bottom: Quentin Schwinn, NASA Glenn Research Center **153** Top: courtesy of GE Aircraft Engines; bottom: courtesy of NASA **154** Bottom left: David Joel Photography, all others, Sam Levitan **155** Sam Levitan **156** Top: courtesy of the Department of Defense; bottom: courtesy of the U.S. Air Force **157** Top left and right: courtesy of the U.S. Navy; bottom: courtesy of the U.S.A.F. Reserves **158** Top: courtesy of NASA; bottom: courtesy of the U.S. Air Force **160** Top: © Burke / Triolo Productions / Getty Images; bottom left: © StockTrek / Getty Images; bottom right: courtesy of NASA **161** Courtesy of NASA **162** Background: courtesy of NASA; bottom right: SWE Archives, Walter P. Reuther Library, Wayne State University **163** Top: courtesy of Pratt & Whitney; bottom: SWE Archives, Walter P. Reuther Library, Wayne State University **164–167** Courtesy of NASA **168** Background: © Corbis; inset: courtesy of NASA **169–170** Courtesy of NASA **171** Bottom left: courtesy of NASA; bottom right: Trotti & Associates, Inc. **172** Top: courtesy of NASA; bottom: courtesy of Cascade Pass, Inc. **173** Courtesy of NASA **174–175** Courtesy of NASA/JPL **176** Top: Library of Congress, Prints & Photographs Division, LC-USZC4-1652; bottom left: © Reuters / Corbis; bottom right: Bryce Flynn Photography Inc / Getty Images **177** Top: © Langevin Jacques / Corbis / Sygma; bottom: courtesy of the U.S. Air Force **178** Library of Congress, Prints & Photographs Division, FSA-OWI Collection, left: LC-DIG-fsac-1a35287, right: LC-DIG-fsac-1a35352 **179** Top: courtesy of the U.S. Air Force; bottom: SWE Archives, Walter P. Reuther Library, Wayne State Univ-ersity, © Boris and Milton—Boston **180** Top: © Fox Photos / Getty Images; bottom: Library of Congress, Prints & Photographs Division, LC-USZ62-32833 **181** Top left: SWE Archives, Walter P. Reuther Library, Wayne State University; top right: used with the permission of The MITRE Corporation, © The MITRE Corporation, all rights reserved **182** Left: courtesy of NASA; right: © Marlin's Photographers **183** Left: © Bettmann / Corbis; right: courtesy of the U.S. Navy **184** Background: © Michael Dunning / Getty Images; bottom left: © Kim Steele / Getty Images; bottom right: courtesy of the U.S. Air Force **185** Top left: courtesy MIT Museum; top right: courtesy of the U.S. Air Force; bottom: © Bettmann / Corbis **186** Top and bottom: courtesy of Northrop Grumman; inset: © CDC / PHIL / Corbis **187** Top: Randy Montoya, Sandia National Laboratories; bottom: courtesy of Lockheed Martin **188** Courtesy of iRobot **189** Top right: Randy Wong, Sandia National Laboratories Staff Photographer **190** Bottom left: R. Eric Alving; all others courtesy of DARPA **191** Top left: Captain Mortensen, U.S.A.F.; top right: Sophia Guerci; bottom left: courtesy of DARPA; bottom right: Arielle Morris **192** Top: © Scott Warren / Getty Images; bottom: © Corbis **195** Top left: © Polk Community College; top right: © Time Life Pictures / Getty Images; bottom: from the Archives of the Seneca Falls Historical Society **196** Top: Joel Greenberg and Wendy Stewart Photography; bottom: Texas A&M University Cushing Library **198** Background: © Corbis; inset: © Bettmann / Corbis **199** Top right and left: Special Collections, University of Maryland Libraries **202** Top: Mark Richards **204** Bottom left: courtesy of Chicago

Historical Society, DN-0005862 (cropped) **205** Top: Wendy Pearce Nelson, Blue Fox Photography **207** Top left: Evan Thomas; bottom right: © Daniel Lainé / Corbis **210** © Hajime Ishizeki / Getty Images **211** © Simon Potter / Getty Images **212** Bottom left: © Royalty-Free / Corbis; bottom center: courtesy of Northrop Grumman; bottom right: David Crenshaw, Crenshaw Photographics **213** Top left: Ford Motor Company; center: Northrop Grumman Newport News